ELECTRIC MACHINERY

ELECTRIC MACHINERY

A. Norton Chaston

Associate Professor
of Electrical Engineering
Brigham Young University
Provo, Utah

A Reston Book
Prentice-Hall
Englewood Cliffs, New Jersey 07632

Library of Congress Cataloging in Publication Data
Chaston, A. Norton
    Electric Machinery

    Bibliography: p.
    Includes index.
    1. Electric machinery. I. Title.
TK2000.C48 1986   621.31'042   84-18182
ISBN 0-8359-1580-8

© 1986 by Prentice-Hall
A Reston Book
Published by Prentice-Hall
A Division of Simon & Schuster, Inc.
Englewood Cliffs, N.J. 07632

10 9 8 7 6 5 4 3 2 1

Printed in the United States of America

CONTENTS

PREFACE

The electric machinery class at most engineering schools has been reduced to a one semester course. Therefore, there has been a serious need for a text to abridge the volumes of information on electric machinery and related topics that have accumulated over a period of 150 years. This digest must be suitable for a one semester course and written for the modern electrical engineering students who specialize in electronics, communication, and computers as well as the mechanical, chemical, and civil engineering students who specify most of the motors used. This text must also take into account the many young faculty members who will be assigned to teach this course and who have limited backgrounds in the subject.

1. A BRIEF DESCRIPTION OF THE BOOK

This text has been written for the engineering student who needs a one semester course in the junior or senior year of study. The prerequisites are an understanding of steady state electrical circuit analysis and single-phase power. It is desirable for the student to have completed an introduction to ferromagnetic circuits as presented in an engineering physics course.

2. SIGNIFICANT FEATURES

This text presents a review of steady state single-phase power, dc ferromagnetic circuits, and an introduction to balanced three-phase circuit theory. In many schools these topics have been deleted from the introductory circuit classes because of the crowded engineering curriculum and are therefore an appropriate part of this text. For those who have completed these topics as part of another course, the first three chapters serve as a review as well as a summary of the notation peculiar to this text. The following are additional features of the text.

a. Sufficient theory is provided so that the student can understand the operation and application of the common types of electric machines.

b. Emphasis is on application rather than design. The student is expected to obtain a knowledge of such features as speed-torque characteristics, starting performance, and speed characteristics, rather than volts per turn, magnetic field patterns, and style of windings for the various types of motors. The performance tests for determining the synchronous reactance of synchronous generators and impedances for the induction machines have been left for the two semester texts.

c. Polyphase power and balanced polyphase circuit theory have been included since many engineering students have limited preparation, if any, regarding these topics.

d. A brief review of steady state ferromagnetic circuit theory is included to supplement the material presented in a beginning physics course.

e. Synchronous machine theory is introduced at the same time as single-phase and three-phase circuit theory. It has been found that this approach allows for added emphasis on ac machines and less emphasis on the dc generator. It should be noted that polyphase rectifiers are available for considerably less cost than a dc motor-generator set and require less maintenance. It has seemed wise, therefore, to reduce the emphasis on dc generators.

f. Since most electronics texts have deleted reference to polyphase rectifiers, a chapter on polyphase rectifier circuits has been included.

g. Following a discussion of synchronous machines, the significant features of single-phase and three-phase transformers are presented.

h. Those who use induction motors need an introduction to machine and circuit protection. It is important for the student to understand why a 700 ampere instantaneous trip breaker may be used with a 125 ampere conductor to supply a 100 ampere induction motor. Therefore, a brief presentation is made regarding fuses, circuit breakers, and starting circuits. This is introduced along with three-phase circuit theory in Chapter 3 and then supplemented at appropriate places throughout the text. The total time required for protection and rulings of the National Electrical Code (NEC) is two lecture hours. The necessary code tables required for design are included in the appendix.

i. Metric units. The only non-SI units used in this text are (1) horsepower as applied to machine size and output, (2) American Wire Gauge (AWG) sizes for conductors, and (3) English sizes for conduit.

j. Problems. Many of these problems have been developed from practical questions received from students or engineers or from observations while doing consulting work. Three types of problems are included in each chapter. There are (1) questions that require essay answers, (2) exercises that drill the student on formulae and basic definitions, and (3) problems that combine principles and ideas.

k. A liberal supply of photos and drawings have been included since it is difficult for the student or teacher to have access to the larger and varied equipment discussed in a machinery course.

l. This text is a suitable introduction for the electric power student who plans to take additional classes in electric machinery and related equipment.

3. ACKNOWLEDGMENTS

Without presenting a separate biographical volume it is difficult to give appropriate recognition to those who have influenced this writing. There are my father and mother. Also, there is my second mother who followed after my first mother's death. There are my teachers of elementary grades through university courses and professional courses and seminars. I wish to recognize my associates at Brigham Young University, General Electric, Southern California Edison, Utah Power and Light, and the City of Los Angeles Department of Water and Power. There are the architects, draftsmen, electricians, and equipment representatives who have assisted with many important questions that form part of one's total education.

I wish to recognize the important work of the authors of the many texts from which I have gleaned understanding. These include Puchstein and Lloyd [C10], Tarboux [C12], Fitzgerald and Kingsley [A14], Matsch [A28], Del Toro [A9], McPherson [A29], and Richardson [A34]. I am grateful to the many students who have influenced this work by the questions they have asked and by their patience through the rough draft manuscripts. Special thanks to Laurie Galvin, Pat Beattie, Bill Long, Evan Nixon, and David Brown for assistance in gathering and preparing data. Thanks is little compensation to Mrs. Opal Machan and department secretaries who have prepared the numerous early manuscript revisions. It was John Fredericksen who entered most the line drawing computer graphics. Dr. J. J. Jonsson (long time department chairman and faculty member at BYU) and J. R. Burnett (formerly with General Electric) have offered much constructive criticism while using the manuscript.

Greg Michael, Alice Cave, Kathryn Mullins, Norma Karlin, and Linda Zuk of the Reston Publishing Company have been most helpful and encouraging in completing this publication.

Finally, I wish to thank my wife Gloria and our five children Larry, Keith, John, Carolyn, Anne and their respective spouses for their moral and editorial support.

ELECTRIC MACHINERY

1

INTRODUCTION AND REVIEW

1.1 HISTORY

The first dc generator (or dynamo as it was then called) was a copper disc rotated between the poles of a horseshoe magnet. Turning the disc produced a direct current in the electrical conductor connected between the edge of the disc and the axle. This was an experiment described by Michael Faraday, an English chemist and physicist, in 1831. He also showed that there was a relationship between electricity, magnetism, and motion. He wrote in his diary on the date of March 26, 1831: "The mutual relation of electricity, magnetism and motion may be represented by three lines at right angles to each other . . . If electricity be determined in one line and motion in another, magnetism will be developed in the third; or if electricity be determined in one line and magnetism in another, motion will occur in the third." [N2]

Equations (1.1) and (1.2) are the mathematical expressions for the principles described by Faraday. The potential at the terminals of the generator is defined by the equation

$$\vec{e} = l \, (\vec{u} \times \vec{B}) \tag{1.1}$$

where $\vec{e}$ is the electrical potential in volts

l is the length in meters of the conductor moving in the magnetic field

$\vec{B}$ is the magnetic flux density in teslas

$\vec{u}$ is the velocity of motion of the conductor in meters per second.

The force on the same machine used as a motor is described by

$$\vec{F} = i\vec{l} \times \vec{B} \tag{1.2}$$

where $\vec{F}$ is the force in newtons
$\quad i$ is the current in amperes
$\quad \vec{l}$ is the conductor length in meters
$\quad \vec{B}$ is the magnetic flux density in teslas.
These equations will be applied to the machines described in later chapters.

Joseph Henry (1797-1878), an American physicist, invented a rocker arm motor in 1829, two years before Faraday made his discoveries in England. However, Henry was slow in publishing his results and Faraday is given credit as the first. Henry was a professor at the Albany Academy (Albany, New York) at the time of this invention. He later taught as a physicist at Princeton University, and eventually became the first director of the Smithsonian Institution in Washington, D.C.

Hippolyte Pixii of Paris produced a direct current generator in 1832 by rotating a horseshoe permanent magnet beneath a pair of fixed coils.

Thomas Davenport (1802-1851) of Brandon, Vermont received U.S. patent No. 132 dated February 25, 1837 for his motor. Lloyd [A24], a twentieth century engineer, wrote this dedication to Davenport:

To the Memory of Three Unsung Martyrs

Thomas Davenport of Brandon, Vermont, who invented the electric motor (U.S. Patent #132, dated 1837) and died penniless as a result.

Emily Davenport, his wife, who gave up her silk wedding dress for insulation on the windings of the first electric motor.

Oliver Davenport, itinerant merchant, who sold his horse and wagon to finance his brother's experiments, thereby putting himself out of business.

Alteneck of the Siemens-Halske Co. of Germany invented a "drum armature" called the Siemens dynamo. The older von Siemens brother, Werner (1816-1892), remained with the German operation. The younger brother Wilhelm Siemens (1823-1883), later knighted as Sir Charles William Siemens, took the Siemens Company electroplating process to England and remained there.

Zenobe Theophile Gramme (1826-1901), a German model maker, built the first efficient industrial dc generator, known as the Gramme Generator, in 1869. At the 1873 Venice Exposition one of Gramme's assistants connected a second generator to a system and found that the second machine started to turn as a motor. This emphasized that an electric machine is bidirectional, i.e., a motor or a generator.

An important use for the dc generator was to supply energy for the electric arc light discovered in 1808 by Sir Humphrey Davy (1778-1828), an English chemist. Paul Jablochkoff (1847-1894), a Russian, lighted the boulevards of Paris with his arc lamps. Charles F. Brush installed a large arc lamp powered by his dc generator in the Wanamakers Department Store in Philadelphia (1878) and in the center of Cleveland, Ohio in 1879.

In October 1878, with the capital arranged for by New York lawyer Grosvenor P. Lowrey, the Thomas Edison Laboratory organized the Edison Electric Light Company to invent an incandescent lamp. Hundreds of experiments followed over a fourteen month period at an expenditure of $40,000. Finally, on October 19, 1879, at the Menlo Park, New Jersey laboratory, the first incandescent lamp began to burn and lasted for just over 40 hours. Three years later on September 4, 1882, Edison started operation of the Pearl Street Station on lower Manhattan Island, which supplied energy for 14,000 lights in 900 buildings [N5, p.46]. The generators were developed by Frances Upton and nicknamed the "long-waisted Mary Anns." The success of this project led them to new business. Four of the corporations formed by Edison to build the Pearl Street Station were consolidated into the Edison General Electric Company with headquarters in Schenectady, New York. Edison returned to his research laboratory and remained only as a member of the board of directors of the Edison General Electric Company.

In the meantime Elihu Thomson had moved to Lynn, Massachusetts and in 1883 formed the Thomson-Houston Electric Company to manufacture dc generators, arc lights, and other electrical equipment [N14]. Thomson was the technical leader and Charles A. Coffin, a former shoe equipment manufacturer, was the business manager. In 1889, Coffin arranged a merger with Brush Electric of Cleveland, Ohio. In 1892, the young Coffin was summoned by J. P. Morgan who suggested he arrange to sell his business to the Edison General Electric Company. The wise young Coffin came away as the first President of the General Electric Company, which resulted from the merger of the Thomson-Houston Electric Company and the Edison General Electric Company. William Stanley (1858-1916) developed a transformer in 1885. Between 1885 and 1888 Stanley was the chief consulting engineer for Westinghouse. Stanley then organized the Stanley Electric Manufacturing Company in Pittsfield, Massachusetts. In 1903 the Stanley Electric Manufacturing Company became the transformer department of the General Electric Company. By 1910 most of the General Electric systems were alternating current.

George Westinghouse (1846-1914), a young man from Central Bridge, New York who had perfected the railroad air brake in 1866, eventually became involved with electrical systems. He became familiar with the ac equipment patents held by Nickola Tesla, a young Yugoslovian immigrant. Westinghouse won the Niagara Falls contract for the ac generation system for the two-phase 33 Hz system. The machines were later rewound for three-phase 60 Hz and were reestablished in service for use during World War II.

While Edison was developing the dc system for the Pearl Street station, Siemens of England was developing the ac system, which was installed at the Grosvenor Square Station. This system was flawed in that the generators would vibrate excessively when connected in parallel, and the first five generators supplied five separate distribution systems. Eventually, the problem was traced to dissimilar voltage waves, and ac generators were designed to produce almost pure sine waves. The ac three-phase system is now the standard system of power distribution.

Through the years many individuals have contributed to the design of efficient electric motors, generators, transformers, and related protective and metering equipment. There are also many control devices that play an important part in high speed printing, computing systems, manufacturing processes, etc. These devices include the selsyn, the stepping motor, and the linear motor.

Approximately 70 percent of all electrical energy generated in the United States is used to drive electric motors. In recent years the need for energy savings has brought a need for improved motors. It is true that large transformers and synchronous generators have efficiencies over 99 percent, and large motors have efficiencies over 90 percent. The high efficiency occurs at about two-thirds of maximum load. However, efficiencies are much less at higher or lower loads. Two systems that claim to be energy saving systems are the NASA motor by Frank Nola and the Wanlass motors by Chris Wanlass.

Before motors can be understood, it is necessary to review magnetic circuits, basic transformers, single-phase circuits, and three-phase circuits.

1.2 SINGLE-PHASE CIRCUITS

This section explains the notations used in this text and includes a basic review of steady state single-phase circuit theory.

A. Review of Single-phase Power

Voltage, current, and power are three principle measurements of an electrical circuit. In the equations, the lower case letters indicate instantaneous quantities and the upper case letters indicate steady-state quantities. The alternating voltage is designated as[1]

$$v(t) = V_m \sin (\omega t + \alpha) \tag{1.3}$$

Steinmetz[2] observed that (1.3) could be represented in the exponential or polar form, called a phasor, by using the Euler form of the voltage equation (1.4).

$$v(t) = V_m \sin (\omega t + \alpha) = \text{Im} \left[V_m e^{j(\omega t + \alpha)} \right] \tag{1.4}$$

$$= \text{Im} \left[V_m \cos (\omega t + \alpha) + j V_m \sin (\omega t + \alpha) \right]$$

where Im means "the imaginary terms of the given expression" and Re means "the real terms of the given expression." In practice, the imaginary expression and the time varying term (ωt) are neglected and the magnitude of (1.4) is expressed in a root-mean-square value (V_{rms}), the effective value (V_{eff}), or simply (V), rather than a maximum value (V_m). Therefore, the instantaneous voltage of (1.3) is expressed in phasor form as

$$\mathbf{V} = V e^{j\alpha} = V \underline{/\alpha} \tag{1.5}$$

Electric current quantities are expressed in a form similar to the voltage quantities. Thus the instantaneous current

$$i(t) = I_m \sin (\omega t + \alpha - \theta) \tag{1.6}$$

is expressed in phasor form

$$\mathbf{I} = I e^{j(\alpha - \theta)} = I \underline{/\alpha - \theta} \tag{1.7}$$

In 1890, electricity was still considered a mysterious energy source, and although the mathematics of dc circuits was understood reasonably well, the mathematics of ac circuits was not. Since there were no

---

[1] Some authors prefer to use the real part of (1.4). Therefore

$$v(t) = \text{Re} \left[V_m \cos (\omega t + \alpha) + j V_m \sin (\omega t + \alpha) \right] = V_m \cos (\omega t + \alpha)$$

The sine version is used in this text to be consistent with that used by most electrical machinery authors.

[2] Charles Proteus Steinmetz, 1865-1923, the hunchbacked genius of German birth, was educated as a mathematician and spent most of his life as a scientist, engineer, and inventor with the General Electric Company, Schenectady, New York.

oscilloscopes or oscillographs, it was difficult to define the magnitude of the alternating wave. A major breakthrough came with the invention of the alternating current ammeter and the use of the rms values. The following indicates the meaning for the rms value.

The average heat dissipated by a pure resistance, R, supplied from a dc source is:

$$P_{ave(dc)} = \frac{1}{T} \int_0^T i(t)^2 R dt = \frac{1}{T} \int_0^T I_{dc}^2 R dt = I_{dc}^2 R \tag{1.8}$$

For an alternating current circuit, the heat dissipated in a pure resistance, R, is:

$$P_{ave(ac)} = \frac{1}{T} \int_0^T i(t)^2 R dt = \frac{R}{T} \int_0^T i(t)^2 dt \tag{1.9}$$

For the same power, (1.8) and (1.9) must be equal so that

$$I_{dc}^2 R = \frac{R}{T} \int_0^T i(t)^2 dt \tag{1.10}$$

From (1.10), solve for I_{dc}. Thus

$$I_{dc} = \left(\frac{1}{T} \int_0^T i(t)^2 dt \right)^{\frac{1}{2}} \tag{1.11}$$

The right hand expression in (1.11) is the effective value or root-mean-square (rms) value of the ac current. Therefore,

$$I = I_{eff} = I_{rms} = \left(\frac{1}{T} \int_0^T i(t)^2 dt \right)^{\frac{1}{2}} \tag{1.12}$$

An alternating electric current measured in rms will dissipate the same heat in a given resistance as the same numerical value of dc current. For example: 10 rms A of ac current and 10 A of dc current will dissipate 500 W in a 5 Ω resistance. The ac meters are calibrated so as to indicate the rms value at a given frequency. The following examples show the procedure for converting between rectangular and polar forms.

Example 1.1

Given a sinusoidal current of $i(t) = 20 \sin (377t + 20°)$, express the current in polar form.

Solution

The polar form is[3]

$$\mathbf{I} = \frac{20}{\sqrt{2}} \, e^{j20°} = 14.1 e^{j20°} = 14.1\underline{/20°}$$

Example 1.2

Given the 60 Hz phasor voltage, $\mathbf{V} = 100\underline{/-15°}$, express the voltage in instantaneous sinusoidal form.

Solution

The instantaneous form is

$$v(t) = \sqrt{2}(100)\sin (377t - 15°) = 141 \sin (377t - 15°)$$

The ratio of the voltage phasor to the current phasor is called *impedance,* Z. Therefore,

$$\mathbf{Z} = \frac{Ve^{j\alpha}}{Ie^{j(\alpha - \theta)}} = \frac{V}{I} \, e^{j\theta} = R + jX \tag{1.13}$$

The unit for impedance is the *ohm* with the real part, R, called *resistance,* and the imaginary part, X, called *reactance.* The reciprocal relationship of current to voltage is called admittance, Y. Therefore,

$$\mathbf{Y} = \frac{Ie^{j(\alpha - \theta)}}{Ve^{j\alpha}} = \frac{I}{Ve^{j\theta}} = G - jB = \frac{1}{\mathbf{Z}} \tag{1.14}$$

The name of the admittance unit is siemens[4] with the real part, G, called conductance and the imaginary part, B, called susceptance.

---

[3] The $\sqrt{2}$ is used to relate the maximum magnitude to the rms magnitude for a sinusoidal waveform.

[4] This unit was once called mhos but has been changed to honor Siemens, a German inventor who improved dc motors.

Example 1.3

Given a voltage source with $v(t) = 141\sin(\omega t + 170°)$ and a load with an impedance of $\mathbf{Z} = 21.3\underline{/61.2°}$, determine the circuit current.

Solution

The phasor voltage is $\mathbf{V} = 100\underline{/170°}$, and the circuit current phasor is

$$\mathbf{I} = \frac{\mathbf{V}}{\mathbf{Z}} = \frac{(100\underline{/170°})}{21.3\underline{/61.2°})} = 4.69\underline{/108.8°}$$

The instantaneous form of the current is

$$i(t) = \sqrt{2}(4.69)\sin(\omega t + 170° - 61.2°) = 6.63\sin(\omega t + 108.8°)$$

Power is defined as the rate at which energy is expended or the rate at which work is accomplished. Thus,

$$p(t) = \frac{dW}{dt} \tag{1.15}$$

For an electrical circuit the instantaneous power, as defined from physics, is

$$p(t) = v(t)\,i(t) \tag{1.16}$$

For the ac system, with $v(t) = V_m \sin \omega t$ and $i(t) = I_m \sin(\omega t + \theta)$

$$p(\omega t) = (V_m \sin \omega t)\,[I_m \sin(\omega t + \theta)]$$

$$= V_m I_m \sin \omega t(\sin \omega t \cos\theta + \cos\omega t \sin\theta)$$

$$= V_m I_m \,[\sin^2 \omega t \cos\theta + (\sin\omega t)(\cos\omega t)(\sin\theta)]$$

$$= V_m I_m \,[\frac{1}{2}(1 - \cos 2\omega t)\cos\theta + (\frac{\sin 2\omega t}{2})(\sin\theta)]$$

$$= \frac{V_m I_m}{2}\,[\cos\theta - (\cos 2\omega t)(\cos\theta) + (\sin 2\omega t)(\sin\theta)]$$

$$= \frac{V_m I_m}{2}\,[\cos\theta - \cos(2\omega t + \theta)]$$

$$p(\omega t) = VI\,[\cos\theta - \cos(2\omega t - \theta)] \tag{1.17}$$

The instantaneous power of a single-phase circuit pulsates at twice the frequency of the supply voltage. Figure 1.1 shows the lagging power factor case. The average power is

$$P_{ave} = \frac{1}{2\pi} \int_0^{2\pi} p(\omega t)d(\omega t) = VI\cos\theta \qquad (1.18)$$

A quantity referred to as the *apparent power,* S, is the product of the voltage and current[5]

$$S = VI \qquad (1.19)$$

A more precise expression for S is the complex form and is defined later in this chapter. The relationship between S and P leads to the power triangle and includes a quantity called *reactive power,* Q. The angle between S and P is called the power factor angle, θ. The related equations are

$$Q = \text{reactive or quadrature power} = VI\sin\theta$$
$$= I^2X = V^2B \qquad (1.20)$$

$$P = \text{real power} = VI\cos\theta = I^2R = V^2G \qquad (1.21)$$

$$\theta = \text{power factor angle}$$

$$\text{power factor} = \frac{P}{VI} = \cos\theta \text{ and expressed in percent or per unit}$$

The "leading power factor" and "lagging power factor" of a circuit element refer to the angle between the voltage and current of that circuit element. Several methods of showing the lagging power factor case, which applies to an inductive (coil) circuit, are shown in Figure 1.1. These include: (a) the plot of the instantaneous voltage and current waves (Figure 1.1b); (b) the plot of the voltage and current phasors with the voltage at some given angle (Figure 1.1d); (c) the plot of the voltage and current phasors with the voltage at zero degrees (Figure 1.1e); (d) the power factor angle as related to the impedance diagrams (Figure 1.1f); and (e) the power factor angle as related to the power triangle (Figure 1.1g).

---

[5] For power triangle quantities, some authors use symbols such as:

P, P_R, or kW for real power with units of W (watts), kW (kilowatts), MW (megawatts), or GW (gigawatt, pronounced jigawatt).

Q, P_X, or kvar for reactive power with units of (volt-ampere-reactive) or vars, kvars, Mvars, and Gvars.

S, P_A, U, MVA, or "volt-amperes" for apparent power with units of VA, kVA, MVA, and GVA.

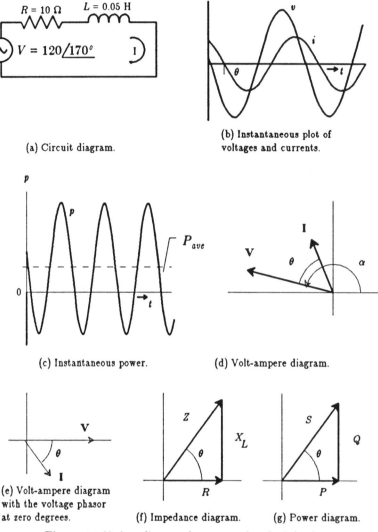

(a) Circuit diagram.

(b) Instantaneous plot of voltages and currents.

(c) Instantaneous power.

(d) Volt-ampere diagram.

(e) Volt-ampere diagram with the voltage phasor at zero degrees.

(f) Impedance diagram.

(g) Power diagram.

Figure 1.1 Various diagrams for representing the inductive reactance circuit of Example 1.4.

Rule. An easy way to remember the lagging power factor is to look at any wall power receptacle and consider the voltage as the reference (normally 120 V) and recall that the current for the inductive load lags behind the source voltage. Hence: lagging power factor.

In describing electrical circuits, three diagrams are commonly used: (1) *phasor diagrams*, Figure 1.1d and 1.1e; (2) *impedance diagrams*, Figure 1.1f; and (3) *power-diagrams*, Figure 1.1g. These three diagrams are separate and distinct, yet are related to the same circuit. The two common quantities between the three diagrams are the power factor angle, θ, and the fact they refer to the same circuit.

Example 1.4

Given the circuit of Figure 1.1a with $\mathbf{V} = 120\underline{/170°}$, determine (a) the impedance diagram, (b) the volt-ampere diagram, and (c) the power diagram for the given circuit.

Solution

(a) Calculate the impedance quantities required for producing the diagram shown in Figure 1.1f.

$$R = 10 \ \Omega$$

$$X_L = 2\pi fL = \omega L = 2\pi(60)(0.05) = 18.85 \ \Omega$$

$$\mathbf{Z} = R + jX_L = 10 + j18.85 = 21.3\underline{/62.1°} \ \Omega$$

$$\theta = 62.1°$$

(b) Sketch the voltage and ampere phasors shown in Figs 1.1d and 1.1e. The voltage was given as

$$\mathbf{V} = 120\underline{/170°}$$

$$\mathbf{I} = \frac{V}{Z} = \frac{120\underline{/170°}}{21.3\underline{/62.1°}} = 5.62\underline{/107.9°} = -1.73 + j5.35 \ \text{A}$$

(c) Sketch the power diagram shown in Figure 1.1g. The power (real power) is a scalar quantity with

$$P = VI\cos\theta = (120)(5.62)\cos 62.1° = 316 \ \text{W}$$

Also, note that $P = I^2R = (5.62)^2(10) = 316$ W.

The *reactive power* is

$$Q = VI\sin \theta = 120 \times 5.62 \times \sin 62.1° = 596 \ \text{vars}$$

Also,

$$Q = I^2X_L = (5.62)^2 18.85 = 596 \ \text{vars}$$

The *apparent power*

$$S = VI = (120)(5.62) = 674 \text{ VA}$$

Also,

$$S = I^2Z = 5.62^2 \times 21.3 = 674 \text{ VA}$$

The power factor is defined as

$$\cos\theta = 0.468 \text{ per unit} = 46.8 \text{ percent lagging}$$

Complex or vector power. The apparent power of (1.19) can be expressed as

$$\mathbf{S} = \mathbf{VI}^* = P + jQ \tag{1.22}$$

where $\mathbf{I}^*$ means the conjugate of $\mathbf{I}$. This is the form preferred by most electric power companies so that lagging power is expressed as a positive quantity. A few electronics-oriented writers, some non-U.S. authors, and some pre-1930 authors prefer using the apparent power equations in the form

$$\mathbf{S} = \mathbf{V}^* \mathbf{I} = P - jQ \tag{1.23}$$

Inductive reactance power is negative for (1.23). The following example compares these two forms and also shows what happens if the conjugate is neglected.

Example 1.5

Given the circuit of Figure 1.1a, calculate the complex **S**

(1) using the form $\mathbf{S} = \mathbf{VI}^*$.

$$\mathbf{S} = (120\underline{/170^\circ})(5.62\underline{/-107.9^\circ}) = 674\underline{/62.1^\circ} = 316 + j596$$

This means a positive reactive power of 596 vars for an inductive load. This is the most common form in use in the United States.

(2) using the form $\mathbf{S} = \mathbf{V}^* \mathbf{I}$.

$$\mathbf{S} = (120\underline{/-170^\circ})(5.62\underline{/107.9^\circ}) = 674\underline{/-62.1^\circ} = 316 - j596$$

This means a negative reactive power of 596 vars for an inductive load. This form is found in some literature but is not the preferred form.

(3) using the incorrect form of $S = VI$.

$$S = (120\underline{/+170°})(5.62\underline{/107.9°})$$

$$= 674\underline{/277.9°} = 93.2 - j668 \; (incorrect)$$

which is an incorrect answer.

Double subscript voltage notation. Often the double subscript notation is easier to use than the single notation method. The notation V_{ab} denotes that a center scale voltmeter with the + terminal connected to point *a* and the – terminal connected to point *b* will deflect upscale if terminal *a* is more positive than terminal *b*. Remember that the modern use of Kirchhoff's Voltage Law (KVL) considers voltage drops as positive and voltage rises as negative. The following examples demonstrate the principle.

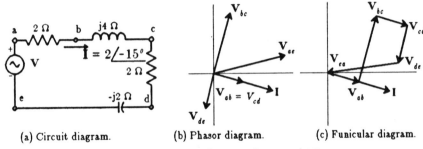

| (a) Circuit diagram. | (b) Phasor diagram. | (c) Funicular diagram. |

Figure 1.2 Diagrams for Examples 1.6 and 1.7.

Example 1.6

Write the voltage equation for the circuit of Figure 1.2a where $I = 2\underline{/-15°}$ A.

$$\mathbf{V}_{ab} + \mathbf{V}_{bc} + \mathbf{V}_{cd} + \mathbf{V}_{de} + \mathbf{V}_{ea} = 0$$

$$2(2\underline{/-15°}) + j4(2\underline{/-15°}) + 2(2\underline{/-15°}) + (-j2)(2\underline{/-15°}) - \mathbf{V}_{ae} = 0$$

$$4\underline{/-15°} + 8\underline{/75°} + 4\underline{/-15°} + 4\underline{/-105°} - \mathbf{V}_{ae} = 0$$

Thus,

$$\mathbf{V}_{ae} = (8.76 + j1.79) = 8.94\underline{/11.57°}$$

Example 1.7

Sketch the voltage phasors for the circuit of Example 1.6 (a) on the real-imaginary axes and (b) funicular diagram. The answers are shown in Figures 1.2b and 1.2c, respectively.

Double subscript current notation. The first letter of the double subscript notation for current designates the source of current and the second letter designates the sink as shown in Example 1.8, which is an application of Kirchhoff's Current Law (KCL).

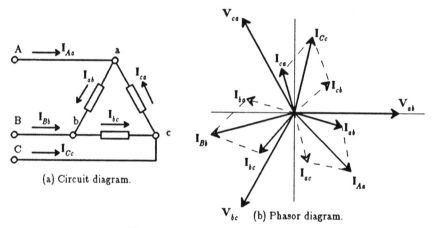

(a) Circuit diagram.   (b) Phasor diagram.

Figure 1.3 Diagram for Example 1.8.

Example 1.8

Write the equations for the line currents of the circuit in Figure 1.3a in terms of the load currents.

$$\mathbf{I}_{Aa} = \mathbf{I}_{ab} + \mathbf{I}_{ac} = \mathbf{I}_{ab} - \mathbf{I}_{ca}$$

$$\mathbf{I}_{Bb} = \mathbf{I}_{ba} + \mathbf{I}_{bc} = -\mathbf{I}_{ab} + \mathbf{I}_{bc}$$

$$\mathbf{I}_{Cc} = \mathbf{I}_{ca} + \mathbf{I}_{cb} = \mathbf{I}_{ca} - \mathbf{I}_{bc}$$

These current phasors are plotted as Figure 1.3b.

1.3 PREVIEW OF ROTATING MACHINES

Two major categories of machines are the direct current and the alternating current machines with several variations within each of these two categories. The following is a list of common machines.

Types of Rotating Machines

I. Polyphase and Direct Current
    A. Polyphase Synchronous
    B. Polyphase Induction
        1. Wound-rotor
        2. Squirrel cage
    C. Direct Current
        1. Permanent magnet
        2. Separately excited
        3. Self-excited
            (a) Shunt
            (b) Series
            (c) Compound
                (1) Cumulative
                (2) Differential

II. Single-phase Motors
    A. Single-phase Induction
        1. Split phase
        2. Capacitor start
        3. Permanent split capacitor (PSC)
        4. Two-value capacitor
        5. Shaded pole
    B. Single-phase Synchronous
        1. Reluctance
        2. Hysteresis
    C. Series (Universal)

A. Synchronous Machines

A simple single-phase synchronous machine similar to the emergency power supply generators available from many mail-order houses consists of a heavy winding in a stationary assembly called a *stator* with an electromagnet mounted on the rotational assembly called a *rotor*. As the rotor is moved a current is induced in the stator winding. This induced stator current varies in magnitude and direction with time, hence the name alternating current (Figure 4.11). If additional stator coils are included, the machine is said to be a polyphase machine with the most common type being the three-phase machine (Figure 4.15).

The synchronous generator is also known by other names such as the *alternator*. Some generators are constructed so as to operate from a slowly driven, vertically mounted hydraulic (hydro) water turbine and are called *hydroelectric-generators* or *hydro-alternators* (Figure 4.1). Other generators are driven by fossil- or nuclear-powered steam turbines (Figures 4.2-6) or fossil-fueled "gas turbines" (Figures 4.7-8) operating at 1800 or 3600 rev/min. These are often called *turbo-alternators* or *turbo-generators*. The IEEE recommended name is the *synchronous generator*.

If an appropriate polyphase voltage is supplied to the stator windings and the synchronous machine brought up to speed, as will be described later, the machine will rotate as a motor at a speed proportional to the frequency of the supply voltage. If the supply frequency is held constant the machine will rotate at a constant or synchronous speed—hence, the name *synchronous motor* (Figure 4.9).

B. Polyphase Induction Machines

The polyphase induction machine is normally used as a motor. The stator assembly is identical to that of the synchronous machine. The rotating assembly, however, consists of coils of conductors. If these conductors are uninsulated and shorted to end rings at each end of the rotor, the machine is called a *squirrel-cage* induction motor. If the rotor windings are insulated and attached to slip rings at one end of the rotor assembly, the machine is referred to as a *wound-rotor* induction motor. The squirrel-cage machine is less expensive to purchase and requires a minimum of maintenance. However, it is essentially a constant speed machine operating at a speed that lags 3 to 5 percent behind the synchronous speed of the stator rotating field. (The rotating field will be explained in Chapter 4.) The wound-rotor machine may be used in many unique ways by connecting resistors, voltage sources, etc., to the rotor slip rings. These features will be mentioned in Chapter 7.

C. dc Machines

A dc generator is essentially the ac synchronous generator with a mechanical rectifier called a *commutator* that converts the ac to dc. In order to effect this mechanical rectification, the current-carrying conductors are mounted on the rotating assembly called the *armature,* and the electromagnet field is mounted on the stationary field assembly. The machine is bidirectional, i.e., it will operate as a motor or as a generator.

The method by which the electrical energy is supplied to the electromagnet field or fields influences the motor or generator operation—hence, the names (1) separately excited, (2) shunt, (3) series, (4) cumulative compound, or (5) differential compound. The features of the dc machine will be presented in Chapter 9.

D. General Comparison of Machines

The dc shunt motors are used when speed variation is desirable, and the dc series motor is used when high starting torque or high speed is required.

The dc machines generally cost more to purchase and require more maintenance than the ac induction motor. The ac induction motor is basically a
constant speed machine with lower initial cost and the lowest maintenance
requirement. In recent years, several novel mechanical and electronic systems have been developed for changing the speed of an induction motor.

The major advantage of the polyphase synchronous machine is the
capability to adjust the power factor. The power factor of a synchronous
motor may be varied by adjusting the dc field. Another advantage is constant speed for a constant frequency.

PROBLEMS

1.1 Convert the following instantaneous expressions to rms phasor
expressions. Plot the resulting phasors.

(a) $v(t) = 163 \sin 377t$

(b) $v(t) = 163 \sin (377t + 25°)$

(c) $v(t) = 163 \sin (377t + 229°)$

(d) $i(t) = 20 \sin (377t - 15°)$

(e) $i(t) = 20 \sin (377t - 115°)$

1.2 Convert the following 60 Hz rms phasors to instantaneous expressions.

(a) $\mathbf{V} = 480\,e^{j120°} = 480\underline{/120°}$

(b) $\mathbf{V} = 230\,e^{j240°} = 230\underline{/240°}$

(c) $\mathbf{I} = 115\,e^{j75°} = 115\underline{/75°}$

1.3 Express the following phasors in polar form.

(a) $5.76 + j3.71$

(b) $17.3 - j5.94$

(c) $-0.0431 - j0.0579$

(d) $-0.0254 + j3.87$

1.4 Express the following phasors in rectangular form.

(a) $6.43\underline{/-18.0°}$

(b) $42.9\,e^{j273.1°}$

(c) $0.712\underline{/-175.2^\circ}$

(d) $13.8e^{-j35.8^\circ}$

1.5 Given the three phasors, perform the indicated calculations.

$$\mathbf{A} = 5.76 + j3.71 \ , \quad \mathbf{B} = 8.91e^{j43.1^\circ} \ , \quad \mathbf{C} = 2.17e^{-j163.2^\circ}$$

(a) $\mathbf{A} - \mathbf{B}$, (b) $\mathbf{B}\,\mathbf{C}$, (c) $\dfrac{\mathbf{C}}{\mathbf{A}}$, (d) $(\mathbf{A})^3$, (e) $\mathbf{C}^{\frac{1}{3}}$

1.6 Express the following phasors in polar form.

(a) $571 + j793$

(b) $0.0691 - j0.0314$

(c) $-2.73 - j8.56$

(d) $-35.2 + j15.9$

1.7 Express the following phasors in rectangular form.

(a) $0.0298e^{j16.9^\circ}$

(b) $0.457e^{-j80.2^\circ}$

(c) $5760\underline{/114.2^\circ}$

(d) $6.31\underline{/-114.9^\circ}$

1.8 Given the two phasors

$$\mathbf{A} = (83.2 + j26.9) \ \text{and} \ \mathbf{B} = 65.1\underline{/-32.9^\circ},$$

perform the following calculations.

(a) $\mathbf{A} - \mathbf{B}$ (b) $\mathbf{A}\,\mathbf{B}$ (c) $\dfrac{\mathbf{B}}{\mathbf{A}}$ (d) $\mathbf{A}^4$ (e) $\mathbf{A}^{\frac{1}{4}}$

1.9 What is the rectangular value of the voltage, V_{ab}?

$$\mathbf{V}_{ab} = (1.0\underline{/120^\circ})(13.2 - j25.9) + (1.0\underline{/-120^\circ})(-21.7 - j10.3) + 8.93\underline{/71.3^\circ}$$

1.10 Given a series RL circuit with a resistance of 16 Ω, an inductance of 0.1 H, and a voltage source of $v(t) = 218 \sin(200t - 31.2^\circ)$, determine the following: (a) phasor current (amperes), (b) power (watts), (c) reactive power (vars), (d) apparent power (volt-amperes), and (e) power factor.

1.11 What is meant by the terms effective and average values? Write the equations that describe the terms.

1.12 Define the terms: power, energy, and work.

1.13 Prove that the effective (rms) value of a sinusoidal voltage defined by the expression $v(t) = V_m \sin\omega t$ is $V = V_{rms} = V_{eff} = V_m/\sqrt{2}$.

1.14 Find the value of voltage (V_{bc}) across the 1 Ω resistor of Figure 1.4.

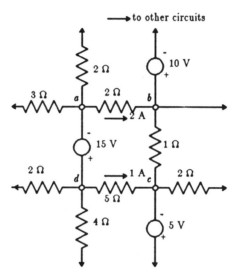

Figure 1.4 Circuit for Problem 1.14.

1.15 A load with an impedance of $\mathbf{Z} = (10.0 + j8.4)$ Ω is connected to a 240 V, 60 Hz main. Determine: (a) average power (P), (b) reactive power (Q), (c) apparent power (S), and (d) load power factor.

1.16 What is the current flow for a 150 W, 120 V incandescent lamp?

1.17 What is meant by power factor?

1.18 A 115 V, 1/3 hp, single-phase induction motor operates at rated load and 60 percent lagging power factor. Assume 60 percent efficiency. The supply voltage is 115 V, 60 Hz. Find the magnitude of the line current.

1.19 Explain the meaning of reactive power, real power, and apparent power.

1.20 A 230 V, 60 Hz, 1725 rev/min, single-phase induction motor operating at full load draws 7.1 A at a lagging power factor of 65 percent. (a) Determine Z, R, X, P, Q, and θ. (b) Sketch the volt-ampere diagram. (c) Sketch the impedance diagram. (d) Sketch the power diagram.

1.21 Find the resultant of the three forces of Figure 1.5 acting on a body with $F_1 = 4\underline{/41°}$, $F_2 = 3\underline{/-32°}$, and $F_3 = 2\underline{/142°}$.

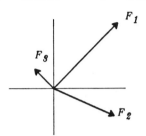

Figure 1.5 Force vectors for Problem 1.21.

1.22 A sewing machine motor is rated at 115 V, single-phase, 60 Hz, and 1.2 A. Find the output available at the shaft (a) in watts and (b) in horsepower, with an input power factor of 85 percent and an efficiency of 70 percent.

1.23 A cooling fan, such as the type used on the Vona 400 minicomputer, is rated at 115 V, 60 Hz, and 14 W input. Find the line current assuming a lagging power factor of 0.87 per unit.

1.24 A cooling fan similar to the ones used on the MBI 063 computer has a motor rating of 115 V and 0.23 A. Assume a power factor of 0.81 per unit. Find the input S, P, Q, Z, R, and X for the motor.

1.25 A motor with an input of 1000 VA at 0.75 lagging power factor is connected in parallel with 1000 W of incandescent lights. What is the total S, P, Q, and pf required by these two loads?

2

MAGNETIC CIRCUITS

Magnetic circuits form part of many of our modern conveniences. For example, the electric clock turns on a radio and wakes you to the gentle music produced by an electromagnetic speaker. As you rise, you switch on electric lights and cook your breakfast on an electric range that receives electrical energy produced by large generators and transmitted economically at high voltages (sometimes as high as 765,000 V) with the aid of transformers (magnetically coupled circuits). You then get in your automobile that starts with an electromagnetic cranking motor and is powered with an internal combustion engine that requires a high voltage produced by a transformer called a spark coil. Are you convinced that magnetic circuits are an extremely important part of your life? In this chapter we will consider the simple ferromagnetic circuit energized by a dc source. However, let us first consider some of the historical events relating to the discovery and application of magnetic principles.

2.1 HISTORY

The first magnets mentioned in recorded history were stones discovered in the ancient country of Magnesia in Asia Minor. These stones were a type of iron ore called magnetite (or loadstone) that many persons believed to be magic. It was discovered that when one of these stones was suspended on a string it would point in a north-south direction; this led to the invention of the compass. The following is a list of some important electrical developments.

1600 — William Gilbert showed the earth was a large magnet.

1819 — Hans Christian Oersted (1777-1871), a Danish physicist and chemist, observed in 1819 that the needle of a compass wavered when placed near a wire carrying current.

1825 — William Sturgeon (1783-1850), an English electrician, discovered the principle of the electromagnet.

1826 — Andre Marie Ampere, the French scientist, discovered the laws of magnetism.

1826 — Joseph Henry, an American physicist published his work "Theory of Electrodynamic Phenomena" in 1826. He showed that parallel conductors carrying electric current in opposite directions repel. He also found coiled wire with an electric current flowing through it acts like a magnet.

1827 — Joseph Henry (1797-1878), an American physicist, became famous for his fundamental discoveries in electromagnetism.

1831 — Michael Faraday (1791-1867), an English chemist and physicist, discovered the principle of electromagnetic induction in 1831. He found that moving a magnet through a coil of wire caused an electric current to flow in the wire. Joseph Henry, an American from upstate New York, independently discovered the same principle shortly before Faraday but failed to publish his discovery.

1873 — James Clark Maxwell (1831-1879), a Scottish scientist, used the discoveries of Faraday to arrive at mathematical descriptions of electrical and magnetic fields. His best known work, "Treatise on Electricity and Magnetism" was published in 1873.

1914 — Robert Andrews Millikan by using the Oil Drop Experiment isolated the electron and existing motor design concepts had to be revised.

2.2 INTRODUCTORY THEORY

The SI units[1] will be emphasized. However, since much of the practical literature still uses other systems of units these are listed in Tables 2.2 and 2.3. If it were not for the nonlinearity of the ferromagnetic cores, the magnetic circuit could well use a set of parameters analogous to electric circuits parameters. Even so, it will be shown that the analogies are extremely helpful in solving the magnetic circuit problems.

---

[1] SI units—the International System of Units (Le Systéme International d'Unités). Given as ANSI standard Z210.1.

A. Magnetic Fields

There is an influence surrounding a conductor carrying an electric current. This influence is called the magnetic flux density, and its direction is described as shown in Figure 2.1.

> **Rule:** Grasp a conductor by the right hand with the thumb extended and pointing in the direction of the current flow. The fingers will curve in the direction of the magnetic field.

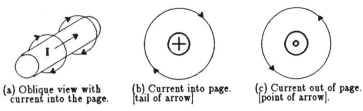

(a) Oblique view with current into the page.

(b) Current into page. [tail of arrow]

(c) Current out of page. [point of arrow].

Figure 2.1 Direction of a magnetic flux density about a conductor.

A corollary to the above rule applies to a coil of wire as shown in Figure 2.2.

> **Rule:** Cup the right hand about a coil with the fingers in the direction of the current flow. The thumb will be pointing in the direction of magnetic flux density.

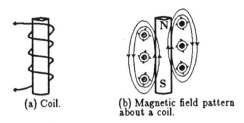

(a) Coil.

(b) Magnetic field pattern about a coil.

Figure 2.2 Direction of magnetic field about a coil.

The direction of flux density is defined by the attraction effect related to a permanent bar magnet. It should be recalled from your classes in physics that one end of a permanent magnet supported so as to rotate freely is attracted toward the north pole of the earth and the opposite end toward the south pole of the earth. These ends of the magnet are called the north and south poles, respectively. It is important to remember that if north poles or south poles (like poles) of two bar magnets are placed close to each other, the bar magnets will repel each other. On the contrary, if a north pole of one bar magnet and a south pole of a second bar magnet (unlike poles) are brought together, the two bars are attracted toward each other.

Rule: Like poles repel; unlike poles attract.

The magnetic field is an influence surrounding a permanent magnet or current carrying conductor and has no origin or termination. However, in order to give meaning to direction, it is said that the field radiates from the north pole of a bar magnet and enters at the south pole.

The field patterns are unaffected by non-magnetic materials such as glass, wood, copper, etc. However, the field pattern may be drastically affected by ferromagnetic materials such as iron. For example, a ferromagnetic core placed in a coil of wire will cause an increase in magnetic field, whereas a wooden core will produce no change in magnetic field.

B. Ferromagnetic Domains

Why do some materials exhibit magnetic characteristics while others do not? The domain theory may be used to answer this question. Magnetism is believed to be of electrical origin. When an electric charge moves with a velocity, a magnetic field is established. In the atom, electrons move in orbits about the heavier nucleus, and, at the same time, each electron, as well as the nucleus, appears to spin about an axis of its own. A spinning electron has a definite angular momentum and a definite magnetic moment. An electron in an atom has, in addition to its spin moments, an angular momentum and a magnetic moment due to its motion in its orbit. The total magnetic moment of the atom is the vector sum of all its component magnetic moments.

The parallel locking of atomic magnetic moments in the crystalline structure extends throughout a limited but somewhat indefinite volume of a ferromagnetic crystal. The reason for this limitation is not completely understood, but experimental evidence shows that, even when an iron crystal as a whole is unmagnetized, tiny neighboring regions called *domains* are completely magnetized. The individual regions, however, have their magnetic moments in different directions and the moments add to zero over the whole crystal.

Any substance that is made up of these spontaneously magnetized and saturated domains is said to possess ferromagnetism. Common substances possessing this characteristic include iron, cobalt, and nickel. Some alloys, called *Heusler alloys,* formed from essentially nonmagnetic materials, have strong magnetic properties. The most highly magnetic of these contain approximately 65 percent copper, 20 percent manganese, and 15 percent aluminum. Conversely, some alloys of iron are not ferromagnetic. For example, some stainless steels are almost nonmagnetic.

The preceding remarks concerning magnetism apply to single ferromagnetic crystals in isotropic[2] assemblies with their crystal axes distributed equally in every direction in an unstrained condition. Most ferromagnetic materials used in engineering are not strain free, nor isotropic. Rolled sheet steel, for example, exhibits widely different magnetic properties for different directions of magnetization, which leads to *grain-oriented* steels. The subject of strain is related to what is called *magnetostriction,* which is the change of physical dimension when magnetic and electric forces in the crystalline structure are disturbed. The magnetostrictive properties of iron are opposite those of nickel. The effect of mechanical tension on iron is to decrease the magnetizing force, whereas the effect of compression is to increase, etc. Thus for iron, the length of the material increases as the magnetization is increased (positive magnetostriction). For nickel, the length decreases (negative magnetostriction).

Strains are produced in materials in several ways: by the presence of impurities (chemical strains), by magnetostriction (latent or residual strain), or by mechanical manipulation (such as cold working).

Ferromagnetic materials are characterized by one or more of the following attributes:

a. They can be magnetized much more easily than other materials.

b. They have a high maximum intrinsic flux density.

c. There is a nonlinear relationship between flux density and the magnetizing force.

d. An increase in magnetizing force produces a change of flux density that is different from the change experienced for a decrease in the magnetizing force (hysteresis).

e. They retain magnetization when the magnetizing force is removed.

---

[2] Isotropic. Exhibiting properties with the same value when measured along axes in all directions.

2.3 PROPERTIES OF PRACTICAL MATERIALS

Of the materials valuable for ferromagnetic applications, iron is the least expensive. It is used in commercially pure form for the frames of many machines. In alloys of 0.25 to 4 percent silicon, it is used for laminations for many alternating current machines. In alloys of nickel, aluminum, and chromium, it is used for permanent magnets. The following is a list of silicon sheet steels that are in common use:

Field-grade sheets contain about 0.25 percent silicon and are also used for small low-priced motors.

Armature-grade sheets contain about 0.5 percent silicon. This grade is rather soft and easily punched. It is used in small motors and generator field poles, armatures, and other devices in which high flux densities are required but core losses are not of great importance.

Electrical-grade sheets contain about 1 percent silicon and are widely used in commercial motors and generators of small and moderate sizes and medium efficiencies; and in power transformers, relays, and other devices designed for intermittent operation.

Motor-grade sheets contain about 2.5 percent silicon and are used in medium-sized motors and generators of good efficiencies, in control apparatus, and in inexpensive communication transformers.

Dynamo-grade sheets contain about 3.5 percent silicon and are used in high efficiency motors and generators, small power-distribution transformers, and communication transformers.

Several *transformer grades* are available. The losses decrease as the silicon content increases. Transformer grade sheets are used primarily for power transformers, communication transformers, large high efficiency generators and motors, and synchronous condensers.

2.4 MAGNETIC CIRCUIT CONCEPTS

A. Basic Expressions

Motors, transformers, etc., all have ferromagnetic cores with electric current-carrying coils or conductors. In order to establish mathematical expressions, the following quantities have been named.

Magnetic flux , ϕ, in webers.

Magnetic flux density, B, in teslas (webers/square meter) where

$$B = \frac{\phi}{A} \qquad (2.1)$$

and A is the cross-sectional area perpendicular to the flux path.

Magnetic field intensity, H, where H is defined by the expression:

$$H = \frac{NI}{l} \qquad (2.2)$$

with NI = ampere-turns; l = mean length of the flux path.

Permeability, μ, is the proportionality between flux density and magnetic field intensity such that:

$$B = \mu H \qquad (2.3)$$

The permeability of free space in metric units has been found to be:

$$\mu_o = 4\pi \times 10^{-7} \qquad (2.4)$$

It is common practice to describe the permeability of a material relative to the permeability of free space and is expressed as *relative permeability, μ_r.* Thus,

$$\mu = \mu_o \mu_r \qquad (2.5)$$

A list of representative relative permeabilities is given in Table 2.1

Table 2.1 Typical Values of Relative Permeability	
Material	μ_r
Paramagnetic Substances	
Air	1.000 038
Aluminum	1.000 023
Diamagnetic Substances	
Copper	0.999 991 2
Water	0.999 991 0

Reluctance, R, represents the opposition to magnetic flux and is related by

$$R = \frac{l}{\mu A} \qquad (2.6)$$

Magnetizing force, F, is also symbolized as

$$F = NI = \text{ampere-turns} = At = \text{mmf} \tag{2.7}$$

The *magnetic circuit law* is defined by the expression:

$$\phi = \frac{F}{R} \tag{2.8}$$

Since the reluctance, R, is nonlinear for ferromagnetic circuits, (2.8) is not always convenient to use. But it does allow a convenient analogy with electric circuit theory that will be shown later. Before introducing the magnetic circuital laws, let us first consider hysteresis losses and eddy currents in order to help in understanding the nonlinear features.

Table 2.2
Units for Magnetic Quantities

Symbol	SI	CGS	English
ϕ	weber	maxwell	lines
B	teslas (webers/m^2)	gauss (maxwells/cm^2)	lines/in^2
μ_o	$4\pi \times 10^{-7}$	1	3.20
H	NI/l	$0.4\pi NI/l$	NI/l
	ampere-turn/meter	oersteds (gilbert/cm)	ampere-turn/inch
F	NI	$0.4\pi NI$	NI
	ampere-turn	gilbert	ampere-turn

B. Hysteresis[3]

When a ferromagnetic material is magnetized by reorienting the magnetic domains, the effect is not fully reversible. When the external influence is removed the magnetic material does not return to its original state. Experimentally this could be found as follows: Wind a coil about a ferromagnetic core, and then measure the flux density for a unidirectional magnetizing force. As shown in Figure 2.3a, at point 0 the flux density would be zero for no magnetizing force for a previously demagnetized system. As the magnetizing force is increased, the flux density will increase along line 0-1. If the magnetizing force is then decreased, the flux density will decrease along line 1-2. The flux that remains at point 2 is referred to as the *residual flux*. If the magnetizing force is reversed in direction, the flux density

---

[3] Hysteresis comes from the greek word "hysterios" meaning to lag behind.

Table 2.3		
Conversion Constants		
Multiply	by	to Obtain
ϕ webers	10^8	lines
webers	10^8	maxwells
B teslas (webers/m^2)	10^4	gauss
	10	kilogauss
	$2.542 \times 10^4 = 6.4516 \times 10^4$	lines/sq. inch
	64.516	kilolines/sq. inch
webers/sq. cm	6.4516×10^8	lines/sq. inch
gauss	10^{-4}	teslas
kilogauss	0.1	teslas
lines/sq. inch	1.550×10^{-5}	teslas
kilolines/sq. inch	0.015 500	teslas
NI ampere-turn	$0.4\pi = 1.25464$	gilberts
H ampere-turns/meter	$2.54/100 = 0.00254$	ampere-turn/inch
	$0.4\pi/100 = 0.012566371$	oersteds
ampere-turns/inch	$0.4\pi/2.54 = 0.494739$	oersteds
	$100/2.54 = 39.370078$	ampere-turns/meter
oersteds (gilberts/cm)	$2.54/(0.4\pi) = 2.021268$	ampere-turns/inch
	$100/(0.4\pi) = 79.577472$	ampere-turns/meter

will decrease along line 2-3 to zero. If the magnetizing force is further decreased, the flux density will increase in the negative direction along line 3-4. As the magnetizing force is increased, the flux density will move along line 4-5 to the negative residual value. Thus, as the magnetizing force is varied between a maximum positive value and a maximum negative value the flux density will follow what is called the *hysteresis loop*. If the maximum value of the magnetizing force is varied, a series of *minor hysteresis* loops will be observed as shown in Figure 2.3b. As the maximum value of the magnetizing force is increased to a large value the area inside the circumscribing curve will remain almost constant. The magnetic material is said to be *saturated*. The curve produced by connecting the ends of the minor loop is called the magnetization curve as shown in Figure 2.3c. Also, magnetization curves are shown for several ferromagnetic materials in Figure 2.4. If the magnetizing force is varied between zero and a given positive value, the flux density will follow the curve as shown in Figure 2.3d, which is referred to as a *partial hysteresis* loop.

Just as a metal strip gets warm when it is repeatedly flexed, a ferromagnetic material gets warm if it is repeatedly magnetized. In both cases this heat is related to the expending of energy. The circumscribed area of the hysteresis loop is proportional to the *coercive* energy or *hysteresis loss*.

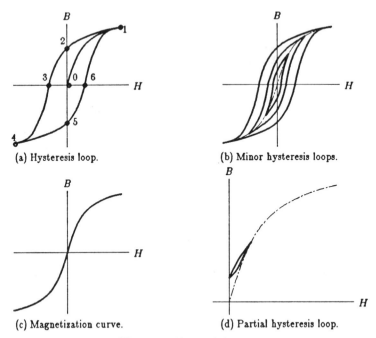

(a) Hysteresis loop.

(b) Minor hysteresis loops.

(c) Magnetization curve.

(d) Partial hysteresis loop.

Figure 2.3 Hysteresis loops.

Magnetic materials used for solenoids, relays, motors, etc. are selected to have a minimum of hysteresis loss while materials used for permanent magnets have a large residual magnetism and a related large hysteresis loss. In some magnetic materials the hysteresis loop is almost rectangular. These materials are well suited for information storage elements in computers. Steinmetz developed an empirical formula for magnetic steels that gives the hysteresis power loss in watts per unit volume as:

$$P_h = K_h B_m^n \tag{2.9}$$

where K_h and the exponent n vary with the core material. Originally, Steinmetz determined that n was approximately 1.6. Since his investigations in 1892, ferromagnetic materials are available with n between 1.5 and 2.5. However, the n is not necessarily constant for a given material indicating that (2.9) is not sufficiently accurate for general use.

If the magnetizing force is produced by an alternating current, the hysteresis loop is traced once each cycle. As the frequency is increased the area inside the curve is also increased. This additional size is caused by eddy currents.

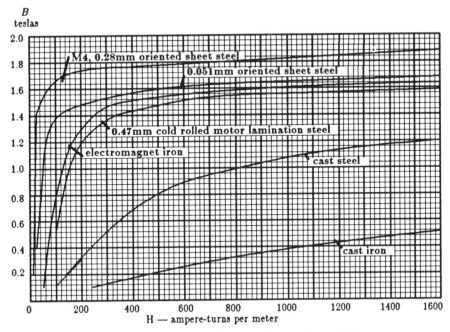

Figure 2.4 dc magnetization curve for several materials.

C. Eddy Currents

An electric current will produce a magnetic field as shown in Figure 2.5. Conversely, a changing magnetic field will produce a changing electric current as shown in Figure 2.5b.

These eddy currents produce I^2R (*or* V^2/R) losses that appear as added heat to the core. Such loss can be reduced by decreasing the effective voltage or increasing the effective resistance. This can be done quite simply by laminating the core so as to reduce the cross-sectional area perpendicular to the path of the flux. These laminations must be electrically insulated from each other. The eddy current power loss is given by the expression:

$$P_e = K_e f^2 B_m^2 \tag{2.10}$$

where K_e depends on the resistivity of the core material and B_m is the maximum value of the sinusoidally varying flux density. Although the topic of the eddy currents has been included in this section, it does not directly apply to the dc systems that will be described in the next few sections.

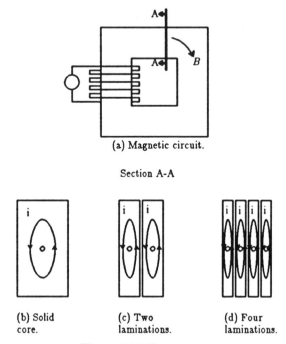

(a) Magnetic circuit.

Section A-A

(b) Solid
core.   (c) Two
                     laminations.   (d) Four
                                           laminations.

Figure 2.5 Eddy currents.

D. Magnetic Circuit Problems

Consider the magnetic circuit that is excited by a dc current. Practical examples of this type of device include an electromagnet for lifting iron and steel, an electric solenoid for starting an automobile, a relay for turning on the furnace in your home, and a variable speed direct current motor. The magnetic circuit problem can be solved in a manner that is analogous with that of electric circuits. For this reason the following two rules are often called "Kirchhoff's law of magnetic circuits."

1. *Magnetic Circuital Laws.* Two magnetic circuit rules may be stated as

> **Rule**: The summation of the mmf drops about a magnetic loop
> equals zero.

$$\sum mmf_{(loop)} = 0 \qquad (2.11)$$

Rule: The summation of the magnetic fluxes at a junction is zero.

$$\sum \phi_{(junction)} = 0 \qquad (2.12)$$

Table 2.4	
Analogous quantities in Magnetic and Electrical Circuits	
Magnetic Circuit Quantity	Electric Circuit Quantity
magnetomotive force F	electromotive force V
magnetic flux ϕ	current I
reluctance R	resistance R
magnetic flux density B	current density J
Permeability μ	Resistivity ρ

Note how these expressions are analogous to Kirchhoff's electric circuit laws which state that the sum of voltage drops about a loop equals zero and the sum of the electric currents at a junction equals zero. This leads to the list of analogies shown in Table 2.4. It is often useful to think in terms of these analogies when solving a magnetic problem.

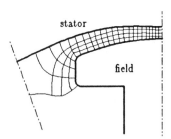

Figure 2.6 Magnetic flux plot for a synchronous motor dc field.

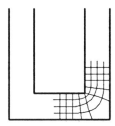

Figure 2.7 Magnetic flux distribution at the corner of a rectangular core.

2. *Air Gap Fringing.* Many ferromagnetic circuits include an air gap. For a large air gap such as that shown in Figure 2.6, it is necessary to make a flux plot. This requires procedures beyond the scope of this text. However, it should be noted that the flux lines leave and enter a ferromagnetic material perpendicular to the iron surface. Even though the flux is the same in the air gap as in the core, the flux density is different. An approximation rule to account for the apparently larger path called fringing is:

Rule: When the length of the air gap is less than 1/10 the width or depth of the ferromagnetic core, the length of the air gap is added to the dimensions of the core to determine the effective cross-section of the air gap.

Example 2.1

A 0.1 mm cut is made in the core of Figure 2.8. What is the effective cross-sectional area of the air gap?

Solution

Add the length of the air gap to each dimension of the core. Thus, the effective area becomes

$$A_{gap} = (20 + 0.1)(25 + 0.1) = 504.5 \text{ mm}^2 = 5.045 \times 10^{-4} \text{ m}^2$$

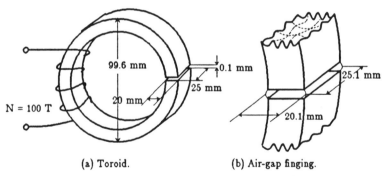

(a) Toroid.   (b) Air-gap finging.

Figure 2.8 Toroidal magnetic circuit for Example 2.1 and Problem 2.13.

3. *Stacking Factor.* The laminations described in the previous section must be electrically insulated from each other. Add to the insulation thickness the air space caused by the uneven and imperfect thickness of each lamination. The ratio of the cross-sectional area of the active ferromagnetic material to the measured overall core area is called the stacking factor.

4. *Examples.* The following four examples outline methods for solving magnetic circuit problems. The first has a constant cross-section core, the second includes an air gap, the third has a core for which the cross-section

varies, and the fourth includes procedures for solving magnetic circuits with parallel paths.

Example 2.2

Given the toroidal core similar to that of Figure 2.8 (except there is no air gap) made of laminations of "0.051 mm, oriented thin sheet steel" and wound with 5 turns of wire. The mean length of the core is 250 mm and has a cross-sectional area of 20 mm by 25 mm. Assume the coil resistance to be 3.4 Ω. Find the magnetic flux density, B, and magnetic flux, ϕ, when a 12 V battery is connected to the coil. Assume a stacking factor of 90 percent.

Solution

The coil current is $I = V/R = 12/3.4 = 3.53$ A. Then the coil ampere-turns is $(NI) = 3.53(5) = 17.7$ At. Solve the magnetic loop equation where $-(NI)_{coil} + (Hl)_{core} = 0$. Then $(Hl)_{core} = (NI)_{coil} = 17.7$ At. Since $l = \pi D$ then $l = \pi(99.6 - 2 \times 10) = 250$ mm $= 0.25$ m. Finally, $H = (Hl)/l = 17.7/0.25 = 70.8$ At/m. From the curve for "0.051 mm, oriented thin sheet steel" of Figure 2.4 the flux density, B, corresponding to a magnetic field intensity, H, of 70.8 At/m is 1.26 T. The area is

$$A = (0.020 \times 0.025)(SF) = (5 \times 10^{-4})(0.9) = 4.5 \times 10^{-4} \text{ m}^2$$

The magnetic flux

$$\phi = BA = (1.26)(4.5 \times 10^{-4}) = 5.67 \times 10^{-4} \text{ Wb}$$

Example 2.3

Make a 0.1 mm wide cut in the toroid of Example 2.2. (a) Calculate the ampere-turns of the coil to produce an air gap flux of 5.67×10^{-4} Wb. (b) What battery voltage would be required to produce the flux specified in step a?

Solution

$$\phi_{air\ gap} = 5.67 \times 10^{-4} \text{ Wb}$$

$$A_{core} = (25)(20)(0.9SF) = 450 \text{ mm}^2$$

$$= 4.5 \times 10^{-4} \text{ m}^2 \text{ effective area}$$

$$A_{air \; gap} = (20 + 0.1)(25 + 0.1)$$

$$= 504.5 \text{ mm}^2 = 5.05 \times 10^{-4} \text{ m}^2 \text{ effective area}$$

$$B_{air} = \frac{\phi_{air}}{A_{air}} = \frac{5.67 \times 10^{-4}}{5.05 \times 10^{-4}} = 1.12 \text{ T}$$

$$H_{air} = \frac{B_{air}}{\mu_o} = \frac{1.12}{4\pi \times 10^{-7}} = 891\,000 \text{ At/m}$$

$$l_{air} = 0.1 \text{ mm} = 10^{-4} \text{ m (as given)}$$

$$(Hl)_{air} = (891\,000)(10^{-4}) = 89.1 \text{ At}$$

$$B_{core} = \frac{\phi_{core}}{A} = \frac{5.67 \times 10^{-4}}{4.50 \times 10^{-4}} = 1.26 \text{ T}$$

From curve on Figure 2.4 for "0.051 mm, oriented thin sheet steel" for $B_{core} = 1.26$, the value of $H_{core} = 70.8$ At/m. Then

$$(Hl)_{core} = 70.8(0.25) = 17.7 \text{ At}$$

Solve the magnetic loop equation.

$$-(NI)_{coil} + (Hl)_{core} + (Hl)_{air} = 0$$

$$(NI)_{coil} = 17.7 + 89.1 = 107 \text{ At}$$

For $N = 5$ turns, the coil current $I = 107/5 = 21.4$ A. This means the required battery voltage is $V = IR = (21.4)(3.4) = 72.6$ V. Note that only 12 V is required to produce a core flux of 5.67×10^{-4} Wb with no air gap. With the addition of the air gap a voltage source of 72.6 V is required to produce the same magnetic flux.

Example 2.4

Given a rectangular core of several types of oriented thin steel sheets with dimensions shown in Figure 2.9, determine the coil current required to produce a flux of 1.30×10^{-3} Wb. The coil has 240 turns. Assume a stacking factor of 0.9.

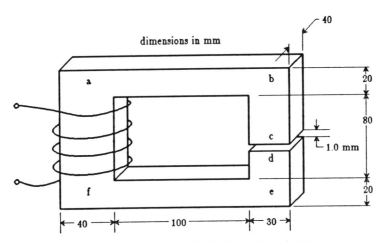

Figure 2.9 Series magnetic circuit for Example 2.4.

Solution

Begin with part *cd* where $\phi = 1.3 \times 10^{-3}$ Wb. The effective cross-sectional area is

$$A_{cd} = (30 + 1)(40 + 1) = 1270 \text{ mm}^2 = 0.00127 \text{ m}^2$$

$$B_{cd} = \phi_{cd}/A_{cd} = \frac{1.30 \times 10^{-3}}{1.27 \times 10^{-3}} = 1.02 \text{ T}$$

$$H_{cd} = \frac{B_{cd}}{\mu_o} = \frac{1.02}{4\pi \times 10^{-7}} = 812\ 000 \text{ At/m}$$

$$H_{cd} l_{cd} = 812\ 000 \times 10^{-3} = 812 \text{ At}$$

Continue with part *ab* where $\phi = 1.3 \times 10^{-3}$ Wb. The effective cross-sectional area is

$$A_{ab} = (20)(40)(0.9) = 720 \text{ mm}^2 = 0.000\ 72 \text{ m}^2$$

The length, l_{ab}, is the mean distance between *a* and *b* of Figure 2.9. This is found to be half of the left section, half of the right section, and the length of the section between. Thus the mean length of l_{ab} is:

$$l_{ab} = (1/2 \times 40 + 100 + 1/2 \times 30) = 135 \text{ mm} = 0.135 \text{ m}$$

$$B_{ab} = \frac{\phi}{A_{ab}} = \frac{1.30 \times 10^{-3}}{0.72 \times 10^{-3}} = 1.81 \text{ T}$$

From the Figure 2.4 curve for "0.28 mm oriented steel"

$$H_{ab} \approx 820 \text{ At/m} \quad \text{for} \quad B_{ab} = 1.81 \text{ T}$$

$$H_{ab}l_{ab} = 820 \times 0.135 = 111 \text{ At}$$

Continue the above procedure for all parts of the magnetic circuit. The results are given in the table below. Then solve the mmf loop equation.

$$-(NI)_{coil} + (NI)_{ab} + (NI)_{bc} + (NI)_{air\ gap} + (NI)_{de} + (NI)_{ef} + (NI)_{fa} = 0$$

$$(NI)_{coil} = 111 + 3 + 812 + 3 + 111 + 4.5 = 1045 \text{ At}$$

$$(NI)_{coil} = \sum(Hl)$$

$$I = \frac{(NI)_{coil}}{N} = \frac{1045}{240} = 4.4 \text{ A}$$

Part	Material	Meas. Area	Eff. Area	Mean Length	ϕ	B	H	Hl
		m^2	m^2	m	Wb	T	At/m	At
cd	Air	0.0012	0.00127	0.001	0.0013	1.02	812,000	812
bc	0.051 mm oriented	0.0012	0.00108	0.0495	0.0013	1.2	60	3.0
de	0.051 mm oriented	0.0012	0.00108	0.0495	0.0013	1.2	60	3.0
ef	0.28 mm oriented	0.0008	0.00072	0.135	0.0013	1.81	820	111
ab	0.28 mm oriented	0.0008	0.00072	0.135	0.0013	1.81	820	111
fa	0.051 mm oriented	0.0016	0.00144	0.10	0.0013	0.903	45	4.5

Example 2.5

Given the cast steel magnetic circuit of Figure 2.10, find the magnitude and direction of the dc current required in coil 2 for a flux, ϕ, of 5×10^{-4} Wb in the direction indicated in the center leg of the core when 2.0 A flows in coil 1. Dimensions and material specifications are included in the table.

Solution

1. Solve left hand loop Hl drops

$$B_{cd} = \frac{\phi_{cd}}{A_{cd}} = \frac{0.0005}{0.00103} = 0.485 \text{ T}$$

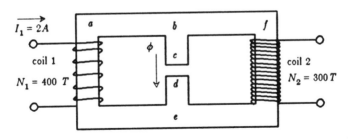

Part	Material	Area m^2	Length m	ϕ Wb	B T	H At/m	Hl At
bc	cast steel	0.001	0.08	0.0005	0.5	295	24
de	cast steel	0.001	0.08	0.0005	0.5	295	24
cd	air gap	0.00103	0.0018	0.0005	0.485	386 000	695
eab	cast steel	0.001	0.45	0.00017	0.17	127	57
bfe	cast steel	0.001	0.40	-0.00033	-0.33	-180	-72

Figure 2.10 Parallel magnetic circuit of Example 2.5.

Continuing with the same procedure calculate the (Hl) drops for l_{bc} and l_{de}. Then

$$H_{cd} = \frac{B}{\mu_o} = \frac{0.485}{4\pi \times 10^{-7}} = 3.86 \times 10^5 \text{ At/m}$$

$$(Hl)_{cd} = (3.86 \times 10^5)(1.8 \times 10^{-3}) = 695 \text{ At}$$

$$(Hl)_{bcde} = 2(24) + 695 = 743 \text{ At}$$

$$(Hl)_{left-loop} = -(NI)_{coil\ 1} + (Hl)_{eab} + (Hl)_{bcde} = 0$$

$$= -400(2) + (Hl)_{eab} + 743 = 0$$

$$(Hl)_{eab} = 800 - 743 = 57 \text{ At}$$

2. Solve for ϕ_{eab}

$$(H)_{eab} = \frac{57}{0.45} = 127 \text{ At/m}$$

From the Figure 2.4 curve for cast steel for $H = 127$, the flux density $B = 0.17$ T. Then

$$\phi_{eab} = (0.17)(10^{-3}) = 1.7 \times 10^{-4} \text{ Wb}$$

3. Solve for the flux at junction b.

$$\phi_{bfe} = \phi_{eab} - \phi_{bcde} = 1.7 \times 10^{-4} - 5 \times 10^{-4} = -3.3 \times 10^{-4} \text{ Wb}$$

This means the magnetic flux is in the *efb* direction.

4. Solve for the mmf drop in leg *bfe*

$$B_{bfe} = \frac{\phi_{bfe}}{A} = \frac{3.3 \times 10^{-4}}{10^{-3}} = 0.33 \ \text{T}$$

From the magnetization curve of Figure 2.4 for "cast steel" with $B = 0.33$ T

$$H_{bfe} = -200 \ \text{At/m}$$

Then

$$(Hl)_{bfe} = -200(0.4) = -80 \ \text{At}$$

5. Solve for the ampere-turns of coil 2

Method 1:

$$(Hl)_{loop \ 2} = -(Hl)_{bcde} + (Hl)_{bfe} + (NI)_{coil \ 2} = 0$$

$$= -743 + (-80) + (NI)_{coil \ 2} = 0$$

$$(NI)_{coil \ 2} = 823 \ \text{At}$$

Method 2:

$$(Hl)_{outside \ loop} = -(NI)_{coil \ 1} + (Hl)_{eab} + (Hl)_{bfe} + (NI)_{coil \ 2} = 0$$

$$-800 + 57 + (-80) + (NI)_{coil \ 2} = 0$$

$$(NI)_{coil \ 2} = 823 \ \text{At}$$

6. Calculate the current in coil 2

$$I = \frac{NI}{N} = \frac{823}{300} = 2.7 \ \text{A}$$

with current *in* at the lower terminal to cause flux in leg *bfe* to flow in direction *efb*.

2.5 ELECTROMAGNETIC FORCE

When a movable ferromagnetic material is placed in a magnetic field, a force exists that causes the movable section to seek a position of minimum reluctance. This is described in the following equation. The derivation for this equation has been given by Fitzgerald (A14(B), p103). In SI units the equation is

$$f = \frac{B_{ag}{}^2 A}{2\mu_o} = \frac{\phi_{ag}^2}{2\mu_o A} \tag{2.13}$$

where f is the force in newtons

B_{ag} is the magnetic flux density in teslas in the air gap

ϕ_{ag} is the magnetic flux in webers in the air gap

A is the cross-sectional area of the air gap in square meters

μ_o is the permeability of free space $= 4\pi \times 10^{-7}$

This equation will provide a means for solving relay and solenoid type problems where the air gap is relatively small.

PROBLEMS

2.1 A coil of 1000 turns is wound on the laminated core of "0.47 mm cold-rolled motor armature steel" shown in Figure 2.11 having a square cross-section of 50 mm on each side (50 mm by 50 mm) and a mean length of 0.5 m. The stacking factor is 0.90. What coil current is required to produce a core flux of 3×10^{-3} Wb? Do not include the air gap in this problem.

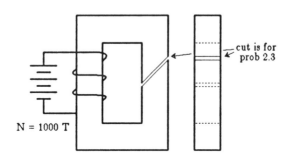

Figure 2.11 Magnetic circuit for Problems 2.1-2.4.

2.2 An air gap 2.5 mm wide is cut perpendicular to the center line of one leg of the core in Problem 2.1. To what value must the coil current be increased in order to maintain the same core flux density? Allowance should be made for air gap fringing.

2.3 A 2.5 mm wide cut is made at an angle of 45 degrees in one leg of the core of Problem 2.1. What value of current is required to produce a core flux of 3×10^{-3} Wb?

2.4 The coil current for the core of Problem 2.2 is adjusted to 2.0 A. What is the magnitude of the magnetic flux in the core? Note: This problem may be solved using the iterative method.

2.5 The "0.051 mm, oriented thin sheet" core of Figure 2.12 has a uniform cross-sectional area of 1200 mm^2 and a mean length of 0.4 m. Coil A has 20 turns and carries 0.5 A, coil B has 40 turns and carries 0.75 A, and coil C carries 1.0 A. How many turns are required for coil C for a flux density in the core of 1.6 T?

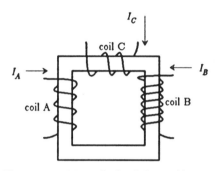

Figure 2.12 Magnetic circuit for Problem 2.5.

2.6 What is the required current in coil 1 of Figure 2.13 so that the magnetic flux density in leg *bed* is 0.3 T when 0.4 A flows in coil 2?

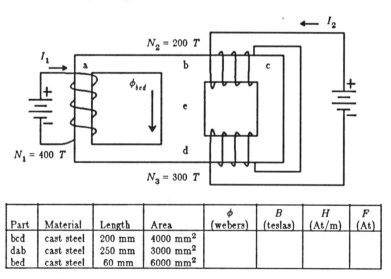

Part	Material	Length	Area	ϕ (webers)	B (teslas)	H (At/m)	F (At)
bcd	cast steel	200 mm	4000 mm^2				
dab	cast steel	250 mm	3000 mm^2				
bed	cast steel	60 mm	6000 mm^2				

Figure 2.13 Magnetic circuit for Problems 2.6, 2.7, and 2.8.

2.7 Repeat Problem 2.6 for a flux density, B_{deb}, of 0.25 T in the up direction.

2.8 Repeat Problem 2.6 for a flux density, B_{deb}, of 0.2 T in the up direction.

2.9 List four examples of magnetic circuits which have not been given in the text.

2.10 Determine the required current in the coil of Figure 2.8 to produce a flux density of 0.8 T in an "electromagnet iron" core. Account for air gap fringing.

2.11 What force in newtons is on the armature of the electromagnet of Figure 2.14 when the flux in the air gap is 1.0×10^{-3} Wb, the area of the air gap is 1000 square mm, and the length of the air gap is 2 mm?

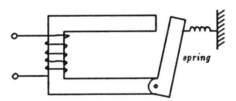

Figure 2.14 Solenoid of Problem 2.11.

2.12 A current of 0.25 A flows in the 1000 turn coil on the cast steel solenoid of Figure 2.15. What is the mechanical force against point A for (a) an air gap of 1.0 mm? (b) almost no air gap?

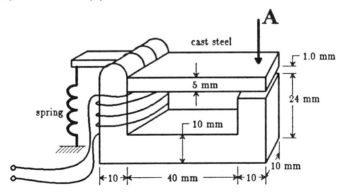

Figure 2.15 Solenoid of Problems 2.12 and 2.13.

2.13 Repeat Problem 2.12 for a "0.051 mm, oriented thin sheet" laminated core with a stacking factor of 0.9.

3

INTRODUCTION TO
THREE-PHASE CIRCUITS

3.1 POLYPHASE CIRCUITS

The common types of polyphase circuits are three-phase for power distribution, two-phase for control motors, and six-phase for power rectifiers and dc power transmission. Four reasons for using three-phase circuits are:

1. A large ac motor requires a rotating magnetic field, which, in turn, requires a three-phase source.

2. There is less voltage drop for a loaded three-phase system than for a single-phase system.

3. There is less line loss (conductor I^2R loss) for a balanced three-phase system because there is no current in the neutral conductor. A single-phase circuit requires a return circuit with associated losses.

4. The instantaneous power is constant.

A. Generation of Three-phase Voltages

A magnetic field moved past a conductor will cause an electric current to flow in a closed loop conductor. Thus, an ideal single-phase generator as shown in Figure 4.11 will produce a single-phase sine wave voltage since there is only one coil or phase. A three-phase system of voltages is produced by mounting three coils as shown in Figure 4.15a. If the coils are arranged so that the maximum of each of the three voltages occurs 120 electrical degrees apart, the system is said to produce a balanced three-phase system. When the voltage of coil *a* leads that of coil *b* and *b* leads *c* and *c* leads *a*, the system is called an *abc* sequence (Figure 3.1a). Also, if the voltage of coil *a* leads *c*, *c* leads *b*, and *b* leads *a*, the system is called *acb* sequence as shown in Figure 3.1c. Some photographs of generators are shown in Figures 4.1 through 4.8. The coils of the generator may be connected in a wye configuration or a delta configuration as shown in Figure 3.2.

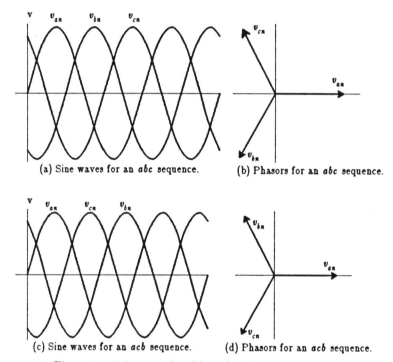

(a) Sine waves for an *abc* sequence.   (b) Phasors for an *abc* sequence.

(c) Sine waves for an *acb* sequence.   (d) Phasors for an *acb* sequence.

Figure 3.1 Voltages produced by a three-phase generator.

B. Circuit Considerations

Calculations relating to unbalanced three-phase circuits require solving the related loop or node equations. For the balanced case the calculations can be simplified by learning some special rules. One simplified form is noted from Figure 3.3a, which shows three single-phase circuits drawn concentrically on the page. How would one solve for the current in any one of the

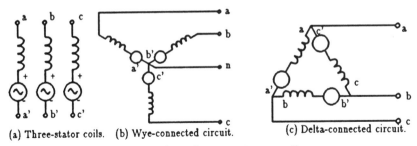

(a) Three-stator coils. (b) Wye-connected circuit. (c) Delta-connected circuit.

Figure 3.2 Three-phase generator connections.

three independent circuits? It becomes obvious that there are three separate and independent problems. Also, it should be obvious that for the balanced three-phase case all the solutions are similar, differing only in the angle relationship. Thus, information obtained for circuit a will provide the magnitude of the voltage and current for circuits b and c.

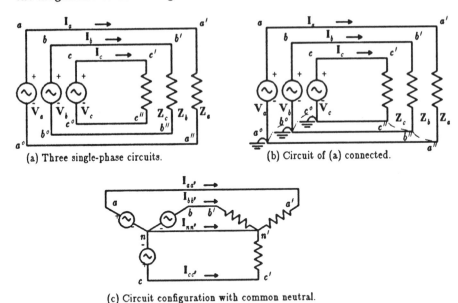

(a) Three single-phase circuits. (b) Circuit of (a) connected.

(c) Circuit configuration with common neutral.

Figure 3.3 Three-phase four-wire wye circuit.

Now consider what happens when the three circuits of Figure 3.3a are connected through a common ground point at a^o, b^o, and c^o as shown in Figure 3.3b. If lines $a^o a''$, $b^o b''$, and $c^o c''$ have zero impedance, then the electrical potential at points a^o, b^o, c^o, a'', b'', and c'' with respect to ground are the same. This means that the three leads could be replaced by one common line called *nn'* and redrawn as shown in Figure 3.3c. This last circuit is known as a four-wire wye circuit. For the balanced case, the current in the common or neutral conductor is zero. This means that the neutral conductor could be removed without affecting the electrical operation of the circuit. The following example should help explain the above ideas.

Example 3.1

Given the circuit of Figure 3.3b with voltage $\mathbf{V}_{aa^o} = 100\underline{/10^o}$ V, $\mathbf{V}_{bb^o} = 100\underline{/250^o}$ V, $\mathbf{V}_{cc^o} = 100\underline{/130^o}$ V, and with a 20 Ω resistor on each of the three loads,

(a) find the currents for Figure 3.3b.

$$\mathbf{I}_{aa'} = \frac{\mathbf{V}_{a'a''}}{\mathbf{Z}_{a'a''}} = \frac{100\underline{/10^o}}{20\underline{/0^o}} = 5\underline{/10^o} \text{ A}$$

$$\mathbf{I}_{bb'} = \frac{\mathbf{V}_{b'b''}}{\mathbf{Z}_{b'b''}} = \frac{100\underline{/-110^o}}{20\underline{/0^o}} = 5\underline{/-110^o} \text{ A}$$

$$\mathbf{I}_{cc'} = \frac{\mathbf{V}_{c'c''}}{\mathbf{Z}_{c'c''}} = \frac{100\underline{/130^o}}{20\underline{/0^o}} = 5\underline{/130^o} \text{ A}$$

(b) find the current $\mathbf{I}_{nn'}$ for Figure 3.3c.

$$\mathbf{I}_{nn'} = -\mathbf{I}_{n'n} = -\left(\mathbf{I}_{aa'} + \mathbf{I}_{bb'} + \mathbf{I}_{cc'}\right)$$

$$= -5\underline{/10^o} + 5\underline{/-110^o} + 5\underline{/130^o} = 0 + j0$$

The above example shows that the rms of the neutral current for the balanced system is zero. Figure 3.4 also shows that the instantaneous value of the neutral current is zero. If the neutral wire is removed, the circuit then becomes the three-wire wye connection of Figure 3.5a. It is surprising how many individuals would shy away from solving a problem such as this and yet never hesitate in attempting to solve the circuit of Figure 3.5b. What is the difference? This emphasizes the advantage of redrawing a circuit if a different form is more easily followed.

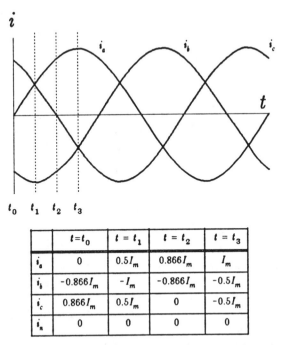

	$t = t_0$	$t = t_1$	$t = t_2$	$t = t_3$
i_a	0	$0.5 I_m$	$0.866 I_m$	I_m
i_b	$-0.866 I_m$	$-I_m$	$-0.866 I_m$	$-0.5 I_m$
i_c	$0.866 I_m$	$0.5 I_m$	0	$-0.5 I_m$
i_n	0	0	0	0

Figure 3.4 Instantaneous currents for a balanced three-phase circuit.

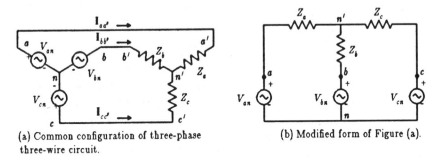

(a) Common configuration of three-phase three-wire circuit.

(b) Modified form of Figure (a).

Figure 3.5 Three-phase three-wire wye circuit.

C. Wye Connected Source

There are six voltages related to the wye circuit of Figure 3.6a. The three voltages V_{an}, V_{bn}, and V_{cn} are related to the generator coils or phases and are called the *phase voltages*. For the wye circuit these are also called the *line-to-neutral* voltages. The voltages V_{ab}, V_{bc}, and V_{ca} are called *line-to-*

line voltages or abbreviated as simply *line voltages*. The line voltages are obtained by phasor addition of the phase voltages.

$$\mathbf{V}_{ab} = \mathbf{V}_{an} + \mathbf{V}_{nb}$$

$$\mathbf{V}_{bc} = \mathbf{V}_{bn} + \mathbf{V}_{nc}$$

$$\mathbf{V}_{ca} = \mathbf{V}_{cn} + \mathbf{V}_{na}$$

This addition for the *abc* sequence is shown graphically in Figure 3.6c. It can be readily shown by trigonometry that the magnitudes of the line voltages are $\sqrt{3}$ times larger than the phase voltages, and the line voltages lead the phase voltages by 30 degrees as shown. Consider the same voltages for the *acb* sequence and note that the magnitudes are the same as for the *abc* sequence but that the line voltages lag the phase voltages by 30 degrees as shown in Figure 3.6e.

There are several cautions in solving three-phase circuits. The phase angles are measured counterclockwise in the same manner as measured in trigonometry for both the *abc* and *acb* sequences. A simple rule of thumb for remembering the *abc* and *acb* sequences is to mentally spin the three phasors of Figure 3.6 in a counterclockwise direction and note that *b* follows *a* and *c* follows *b* for the *abc* sequence. For the *acb* sequence *c* follows *a* and *b* follows *c*. Also, do not resort to the erroneous practice of simply changing the subscripts of $\mathbf{V}_{ab}$ to $\mathbf{V}_{ba}$ and assume that the phase sequence has been changed.

Example 3.2

Given a 208 V, three-phase, wye connected source similar to that of Figure 3.6, select the angle of $\mathbf{V}_{an}$ at zero degrees. Determine the magnitude and angles of all phase and line voltages for the *abc* and *acb* sequences.

Solution

A 208 V source implies a line-to-line magnitude of 208 V. Thus, the phase voltages are $208/\sqrt{3} = 120$ V. Since $\mathbf{V}_{an}$ is given at 0 degrees, the phase voltages for the *abc* sequence are:

$$\mathbf{V}_{an} = 120\underline{/0^\circ}\ \text{V}$$

$$\mathbf{V}_{bn} = 120\underline{/240^\circ}\ \text{V}$$

$$\mathbf{V}_{cn} = 120\underline{/120^\circ}\ \text{V}$$

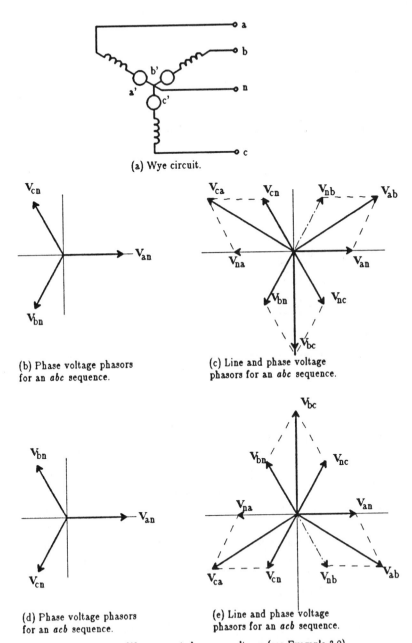

(a) Wye circuit.

(b) Phase voltage phasors for an *abc* sequence.

(c) Line and phase voltage phasors for an *abc* sequence.

(d) Phase voltage phasors for an *acb* sequence.

(e) Line and phase voltage phasors for an *acb* sequence.

Figure 3.6 Wye-connected source voltages (see Example 3.2).

Then the line voltages are calculated to be

$$\mathbf{V}_{ab} = \mathbf{V}_{an} + \mathbf{V}_{nb} = 208\underline{/30^{\circ}} \text{ V}$$

$$\mathbf{V}_{bc} = \mathbf{V}_{bn} + \mathbf{V}_{nc} = 208\underline{/270^{\circ}} \text{ V}$$

$$\mathbf{V}_{ca} = \mathbf{V}_{cn} + \mathbf{V}_{na} = 208\underline{/150^{\circ}} \text{ V}$$

These voltages are shown in Figure 3.6b and c.

For the *acb* sequence, the phase (or line-to-neutral) voltages are

$$\mathbf{V}_{an} = 120\underline{/0^{\circ}} \text{ V}$$

$$\mathbf{V}_{bn} = 120\underline{/120^{\circ}} \text{ V}$$

$$\mathbf{V}_{cn} = 120\underline{/-120^{\circ}} \text{ V}$$

Then the line-to-line voltages are

$$\mathbf{V}_{ab} = 208\underline{/-30^{\circ}} \text{ V}$$

$$\mathbf{V}_{bc} = 208\underline{/90^{\circ}} \text{ V}$$

$$\mathbf{V}_{ca} = 208\underline{/-150^{\circ}} \text{ V}$$

These phasors are shown in Figure 3.6.d and e.

D. Wye Connected Load

At this point, consider the funicular diagram of Figure 3.7b. The funicular diagram is obtained by extending the phasors end to end in the same manner that mechanical forces are sometimes added end on end. Thus, begin with $\mathbf{V}_{ab}$ and add $\mathbf{V}_{bc}$, then complete the triangle by adding $\mathbf{V}_{ca}$. Since the source voltages are balanced and must add to zero, the funicular diagram appears as an equilateral triangle. For the balanced case, the circuit neutral *n* will be the centroid of the triangle. This allows the easy determination of the phase voltages $\mathbf{V}_{an}$, $\mathbf{V}_{bn}$, and $\mathbf{V}_{cn}$. Now transfer the line and phase voltage phasors to the phasor diagram as shown in Figure 3.7c. Next determine the phase currents by considering the phase voltages and phase impedances. The following example should help clarify the procedure.

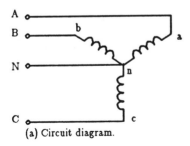

(a) Circuit diagram.

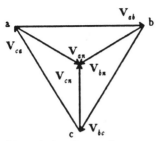

(b) Funicular diagram for a
balanced wye circuit for
an *abc* sequence.

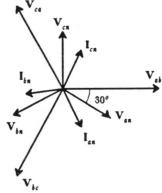

(c) Phasor diagram for a
balanced wye circuit for
an *abc* sequence.

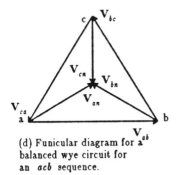

(d) Funicular diagram for a
balanced wye circuit for
an *acb* sequence.

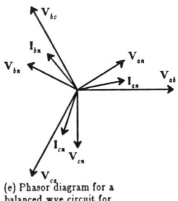

(e) Phasor diagram for a
balanced wye circuit for
an *acb* sequence.

Figure 3.7 Wye-connected load voltages.

Example 3.3

Given a 208 V 60 Hz circuit with a wye load of $(5 + j2)$ Ω per phase similar to that of Figure 3.7, show the voltage and current phasors for the *abc* and *acb* sequences.

Solution

Select the angle for $\mathbf{V}_{ab}$ to be zero degrees. Then the line voltages are:

$$\mathbf{V}_{ab} = 208\underline{/0^\circ} \text{ V}$$

$$\mathbf{V}_{bc} = 208\underline{/-120^\circ} \text{ V}$$

$$\mathbf{V}_{ca} = 208\underline{/120^\circ} \text{ V}$$

The phase voltage magnitude is $208/\sqrt{3} = 120$ V. From the funicular diagram, Figure 3.7b, observe the angles for the phase voltages so that

$$\mathbf{V}_{an} = 120\underline{/-30^\circ} \text{ V}$$

$$\mathbf{V}_{bn} = 120\underline{/-150^\circ} \text{ V}$$

$$\mathbf{V}_{cn} = 120\underline{/90^\circ} \text{ V}$$

The phase current is

$$\mathbf{I}_{an} = \frac{\mathbf{V}_{an}}{\mathbf{Z}_{an}} = \frac{120\underline{/-30^\circ}}{(5 + j2)} = 22.3\underline{/-51.80^\circ} \text{ A}$$

The line current is the phase current for a wye circuit so that

$$\mathbf{I}_{Aa} = \mathbf{I}_{an} = 22.3\underline{/-51.80^\circ} \text{ A}$$

Then

$$\mathbf{I}_{Bb} = \mathbf{I}_{bn} = 22.3\underline{/-171.80^\circ} \text{ A}$$

$$\mathbf{I}_{Cc} = \mathbf{I}_{cn} = 22.3\underline{/68.20^\circ} \text{ A}$$

For the *acb* sequence the values are

$$\mathbf{V}_{ab} = 208\underline{/0^\circ} \text{ V} \quad \mathbf{V}_{bc} = 208\underline{/120^\circ} \text{ V} \quad \mathbf{V}_{ca} = 208\underline{/-120^\circ} \text{ V}$$

$$\mathbf{V}_{an} = 120\underline{/30^\circ} \text{ V} \quad \mathbf{V}_{bn} = 120\underline{/150^\circ} \text{ V} \quad \mathbf{V}_{cn} = 120\underline{/-90^\circ} \text{ V}$$

$$\mathbf{I}_{an} = 22.3\underline{/8.20°} \text{ A} \qquad \mathbf{I}_{bn} = 22.3\underline{/128.2°} \text{ A} \qquad \mathbf{I}_{cn} = 22.3\underline{/-111.80°} \text{ A}$$

$$\mathbf{I}_{Aa} = \mathbf{I}_{an} \qquad\qquad \mathbf{I}_{Bb} = \mathbf{I}_{bn} \qquad\qquad \mathbf{I}_{Cc} = \mathbf{I}_{cn}$$

Where $\quad \mathbf{I}_{an} = \dfrac{\mathbf{V}_{an}}{\mathbf{Z}_{an}} = \dfrac{120\underline{/30°}}{5.39\underline{/21.80°}} = 22.3\underline{/8.20°} A \quad$ A

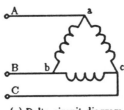

(a) Delta circuit diagram.

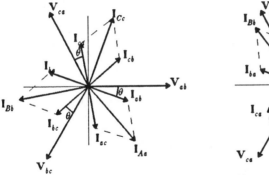

(b) Phasor diagram for a balanced
delta circuit for *abc* sequence.

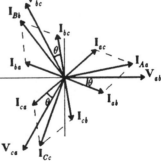

(c) Phasor diagram for a balanced
delta circuit for *acb* sequence.

Figure 3.8 Delta-connected load voltages and currents.

E. Delta-connected Load

For the delta load the phase voltage is the line-to-line voltage. The phase current is determined from the ratio of the given phase voltage and phase impedance. The line currents are then the sum of the appropriate phase currents. Thus, the line currents in Figure 3.8a are

$$\mathbf{I}_{Aa} = \mathbf{I}_{ab} + \mathbf{I}_{ac} = \mathbf{I}_{ab} - \mathbf{I}_{ca}$$

$$\mathbf{I}_{Bb} = \mathbf{I}_{ba} + \mathbf{I}_{bc} = -\mathbf{I}_{ab} + \mathbf{I}_{bc}$$

$$I_{Cc} = I_{ca} + I_{cb} = I_{ca} - I_{bc}$$

Consider the following example for more details.

Example 3.4

Given a 208 V, 60 Hz circuit with a delta load of $(5+j2)$ Ω per phase, similar to that of Figure 3.8, show the voltage and current phasors for the *abc* and *acb* sequences.

Solution

The phase voltages are the line-to-line voltages and are:

$$V_{ab} = V_{AB} = 208\underline{/0^o}\ V$$

$$V_{bc} = V_{BC} = 208\underline{/-120^o}\ V$$

$$V_{ca} = V_{CA} = 208\underline{/120^o}\ V$$

The phase impedance is $(5 + j2) = 5.39\underline{/21.80^o}$ Ω

The phase currents are:

$$I_{ab} = \frac{V_{ab}}{Z_{ab}} = \frac{208\underline{/0^o}}{5 + j2} = \frac{208\underline{/0^o}}{5.39\underline{/21.80^o}} = 38.6\underline{/-21.80^o}\ A$$

$$I_{bc} = \frac{208\underline{/-120^o}}{5.39\underline{/21.80^o}} = 38.6\underline{/-141.80^o}\ A$$

$$I_{ca} = \frac{208\underline{/120^o}}{5.39\underline{/21.80^o}} = 38.6\underline{/98.20^o}\ A$$

The line currents are:

$$I_{Aa} = I_{ab} + I_{ac}$$
$$= 38.6\underline{/-21.80^o} - 38.6\underline{/98.20^o}\ A$$
$$= 41.35 - j52.54 = 66.86\underline{/-51.80^o}\ A$$

$$I_{Bb} = 66.86\underline{/-171.80^o}\ A$$

$$I_{Cc} = 66.86\underline{/68.20^o}\ A$$

For the *acb* sequence the voltages and currents are:

$$\mathbf{V}_{ab} = 208\underline{/0^\circ}\ \mathrm{V} \qquad \mathbf{V}_{bc} = 208\underline{/120^\circ}\ \mathrm{V} \qquad \mathbf{V}_{ca} = 208\underline{/-120^\circ}\ \mathrm{V}$$

$$\mathbf{I}_{ab} = 38.6\underline{/-21.80^\circ}\ \mathrm{A} \qquad \mathbf{I}_{bc} = 38.6\underline{/98.20^\circ}\ \mathrm{A} \qquad \mathbf{I}_{ca} = 38.6\underline{/-141.80^\circ}\ \mathrm{A}$$

$$\mathbf{I}_{Aa} = 66.86\underline{/8.20^\circ}\ \mathrm{A} \qquad \mathbf{I}_{Bb} = 66.86\underline{/128.20^\circ}\ \mathrm{A} \qquad \mathbf{I}_{Cc} = 66.86\underline{/-111.80^\circ}\ \mathrm{A}$$

These are shown in Figure 3.8.

The following is a summary showing the relationship between phase and line quantities for wye and delta circuits:

Wye Load

$$V_{line} = \sqrt{3}\ V_{phase}$$

$$I_{line} = I_{phase}$$

Delta Load

$$V_{line} = V_{phase}$$

$$I_{line} = \sqrt{3}\ I_{phase} \qquad (3.4)$$

F. Power for a Balanced Three-phase Circuit

Since power is a scalar quantity the power delivered to a three-phase load is the sum of the power delivered to each of the three phases. Thus,

$$P_{total} = P_{phase\ 1} + P_{phase\ 2} + P_{phase\ 3}$$

For a balanced three-phase circuit the total power delivered to the load is

$$P_{total} = 3P_{phase} = 3V_{ph}I_{ph}\cos\theta_{ph}$$

For the wye circuit

$$P_{wye} = 3\left(\frac{V_L}{\sqrt{3}}\right)I_L\cos\theta_{ph} = \sqrt{3}V_LI_L\cos\theta_{ph} \qquad (3.6)$$

For a delta circuit

$$P_{delta} = 3V_L\left(\frac{I_L}{\sqrt{3}}\right)\cos\theta_{ph} = \sqrt{3}V_LI_L\cos\theta_{ph} \qquad (3.7)$$

Thus the average power to a three-phase wye or delta load may be determined from the line voltage and line current as given by the expression

$$P = \sqrt{3}V_LI_L\cos\theta_{ph} \qquad (3.8)$$

This is a particularly valuable equation because there is often no convenient way to measure the phase quantities for a totally enclosed machine.

Closely related to the real power are the reactive and apparent power quantities for the balanced case. The three-phase quantities are treated as totals in the same manner as single-phase quantities.

$$S = \sqrt{3} V_L I_L \tag{3.9}$$

$$P = \sqrt{3} V_L I_L \cos\theta_{ph} = S \cos\theta_{ph} \tag{3.10}$$

$$Q = \sqrt{3} V_L I_L \sin\theta_{ph} = S \sin\theta_{ph} \tag{3.11}$$

θ_{ph} = power factor angle

= phase impedance angle[1]

power factor = $\cos\theta_{ph}$

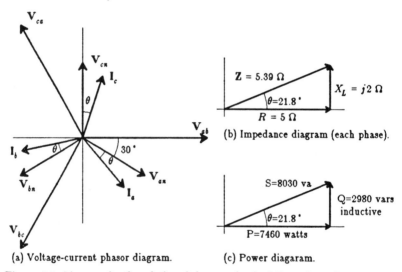

(a) Voltage-current phasor diagram.

(b) Impedance diagram (each phase).

(c) Power diagaram.

Figure 3.9 Diagrams for the solution of the wye circuit of Example 3.5 Part a.

Example 3.5

Polyphase power. Solve for the power delivered to the wye load in Example 3.3 and the delta load in Example 3.4. Also, show the impedance triangle for these examples.

---

[1]Note that the power factor angle refers to the angle between the phase voltage and phase current and is not the angle between the line-to-line voltage and line current.

Solution

(a) For the wye circuit of Example 3.3

$$S = \sqrt{3}\,V_L I_L = \sqrt{3}(208)(22.3) = 8030 \text{ VA}$$

$$\theta_{ph} = 21.80$$

$$pf = 0.93$$

$$P = S \cos\theta_{ph} = 8030 \cos 21.80 = 7460 \text{ W}$$

$$Q = S \sin\theta_{ph} = 8020 \sin 21.80 = 2980 \text{ vars inductive}$$

The results of (a) are plotted in Figure 3.9c.

(b) For the delta circuit of Example 3.4

$$S = \sqrt{3}\,V_L I_L = \sqrt{3}(208)(66.9) = 24{,}100 \text{ VA}$$

$$\theta_{ph} = 21.80^\circ$$

$$pf = 0.93 \text{ per unit power factor}$$

$$P = S \cos\theta_{ph} = 24{,}100 \cos 21.80 = 22{,}400 \text{ W}$$

$$Q = S \sin\theta_{ph} = 24{,}100 \sin 21.80 = 8950 \text{ vars inductive}$$

(c) The impedance per phase for either example is

$$(5 + j2) = 5.39\underline{/21.80^\circ} \ \Omega$$

G. Parallel Loads

Engineers involved in supplying or using electrical energy are concerned about calculating the total electrical load. These loads may be easily determined by combining the real and reactive components of the load. In other words, "watch your P's and Q's." Example 3.6 indicates the procedure for solving parallel loads.

Example 3.6

Parallel connection of three-phase loads. Consider the following 2300 V, 60 Hz loads.

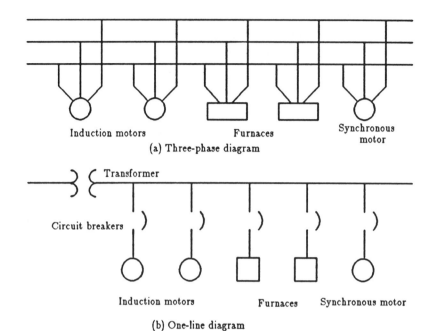

(a) Three-phase diagram

(b) One-line diagram

Figure 3.10 Circuit for Example 3.6.

Induction motors	3200 kVA	60 percent lagging pf
Electrical furnaces	1500 kVA	unity power factor
Synchronous motor	800 kVA	30 percent leading pf

a. What is the total kVA and power factor? See Figure 3.10.

b. What is the magnitude of the line current? Construct a table and solve for the kVA and power factor.

S (kVA)	P (kW)	Q (kvar)	θ (Degrees)	$\cos\theta$ (pf)
3200	1920	+2560	53.1° lag	0.6 lag
1500	1500	0	0	1.0
800	240	−763	72.5° lead	0.3 lead
×	3660	+1797	×	×
4077	×	×	26.2° lag	0.898 lag

Therefore,

S = 4077 kVA at 89.8 percent lagging power factor.

The line current from (3.9) is

$$I = \frac{S}{\sqrt{3}\ V} = \frac{4\ 077\ 000}{\sqrt{3}\times 2300} = 1023\ \text{A}$$

H. Power Factor Improvement

The first reason for keeping the power factor of a system close to unity is increased system efficiency. For a given system voltage, the closer the power factor is to unity, the smaller the current. Since the system electrical losses are the I^2R losses and are related to the square of the current, the less the current, the better the power factor and the more efficient the system.

A second reason for keeping the power factor close to unity is related to the size of the equipment. For a given voltage the apparent power in kVA or (MVA) rating of a transformer, generator, etc., must be larger for a lower system power factor. For example, a 100 MW, unity power factor load requires a 100 MVA generator whereas a 100 MW, 50 percent power factor load requires a 200 MVA generator. The 200 MVA generator will cost approximately twice as much as a 100 MVA generator.

Because of increased losses and equipment size for low power factor systems, the electric utility company supplying energy is permitted to charge a penalty for low power factor loads. Generally, where there is the power factor penalty charge there is an immediate cash savings resulting from the installation of power factor correction devices. Since most practical loads are inductive, the power factor improvement is accomplished by connecting capacitors in parallel with the load (Figure 3.11). For large installations, power capacitors are connected in series and/or parallel combinations. In improving the power factor, the economic principle of diminishing returns applies, and the power factor is seldom corrected to values greater than 95 percent.

A synchronous motor operated at a leading power factor (described in Chapter 4) can be used for power factor correction. A special type of synchronous motor used exclusively for power factor correction is called a *synchronous capacitor* or *synchronous condenser* and is a synchronous motor without a shaft extension.

In summary, a considerable effort is made by electric energy suppliers and users to improve the system power factor. The supplier is able to use smaller (less expensive) supply equipment at an increased efficiency, and the user gets a better energy billing rate.

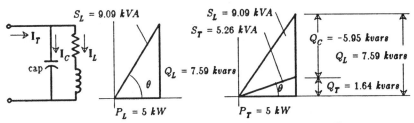

(a) Circuit diagram. (b) Load power diagram.   (c) Combined power diagram.

Figure 3.11 Power factor correction Example 3.7.

Example 3.7

Single-phase power factor correction. Given a 5.0 kW, 115 V, single-phase, 60 Hz, 55 percent lagging power factor load, determine (a) S, P, Q, and I for the given load, (b) S, P, Q, and pf for the combined load, (c) the capacitance in kvars that must be connected in parallel with the load of part (a) to correct the power factor to 95 percent, and (d) size of capacitor of (c) in kVA, ohms, and microfarads (See Figure 3.11).

Solution

(a) $P = 5.0$ kW

$\theta = \arccos(0.55) = 56.63°$

$Q = P \tan\theta = 7.59$ kvar

$S = \dfrac{P}{\cos\theta} = 9.09$ kVA

$I_L = \dfrac{P}{V\cos\theta} = \dfrac{5000}{115 \times 0.55} = 79.1$ A

(b) The power factor corrected load has a total angle of

$\theta_T = \arccos(0.95) = 18.19°$

$pf = \cos\theta =$ given as 0.95 per unit or 95%

$P_T =$ remains the same as for part (a) $= 5.0$ kW

$Q_T = P_T \tan\theta_T = 1.643$ kvars

$$S_T = \frac{P_T}{\cos\theta} = 5.263 \text{ kVA}$$

$$I_T = \frac{P_T}{V\cos\theta_T} = \frac{5000}{115 \times 0.95} = 45.77 \text{ A}$$

since $S_c = Q_c = V_c I_c$, then

$$I_C = \frac{Q_C}{V} = \frac{5949}{115} = 51.73 \text{ A}$$

(c) $Q_c = Q_T - Q_L = 1.64 - 7.59 = -5.95$ kvars

Note that large capacitors or capacitor banks are rated in kvars and kV, thus the size capacitor would be the next standard size larger than 5.95 kvars and 115 V.

(d) At times it is interesting to know the capacitor size in farads (microfarads). Thus,

$$Q_C = \frac{V^2}{X_C} \quad \text{or} \quad X_C = \frac{V^2}{Q_C} = 2.223 \ \Omega$$

or $X_C = -j2.223 \ \Omega$

Since $X_C = \frac{1}{\omega C}$

or $C = \frac{1}{\omega X_C} = \frac{1}{(2\pi 60)(2.223)} = 1193 \ \mu f$

The procedure for solving three-phase problems is similar to that for the single-phase case. The following example should clarify the procedure.

Example 3.8

Three-phase power factor correction. Given a 200 hp, three-phase, 460 V, 60 Hz, squirrel-cage induction motor with a full load current of 240 A at 0.84 per unit power factor (where per unit power factor is the cosine of the power factor angle), (a) determine the size of the capacitor bank (in kvars) that must be connected in parallel with the motor in order to correct the power factor to 0.96 per unit, (b) determine the voltage rating and size of capacitors in ohms and microfarads for a wye connection, and

(c) determine the voltage rating and size of capacitors in ohms and microfarads for a delta connection.

Solution

The motor power quantities are

$$S_M = \sqrt{3}V_L I_1 = \sqrt{3}(460)(240) = 191 \text{ kVA}$$

$$P_M = S \cos\theta_{ph} = 191(0.84) = 161 \text{ kW}$$

$$\theta_{ph} = \arccos 0.84 = 32.86° \text{ lagging}$$

$$Q_M = 191 \sin 32.86° = 104 \text{ kvars}$$

If the capacitor bank is assumed to have no losses, then the compensated system power quantities are

$$P_T = P_M = 161 \text{ kW}$$

$$\theta_T = \arccos 0.96 = 16.26° \text{ lagging}$$

$$Q_T = P_T \tan\theta_T = 161 \tan 16.26° = 47.0 \text{ kvars}$$

$$S_T = \frac{P_T}{\cos\theta_T} = \frac{161}{\cos 16.26°} = 168 \text{ kVA}$$

The capacitor must be sized at

$$Q_C = Q_T - Q_M = 47.0 - 104 = -57 \text{ kvars} = 57 \text{ kvars capacitive}$$

(b) For a wye-connected capacitor bank

$$Q_{phase} = Q_{Cph} = \frac{Q_T}{3} = \frac{V_{ph}^2}{X_{Cph}}$$

$$X_{Cph} = \frac{V_{ph}^2}{Q_{Cph}} = \frac{(V_L/\sqrt{3})^2}{Q_C/3} = \frac{V_L^2}{Q_C}$$

$$= \frac{(460)^2}{57000} = 3.71 \ \Omega$$

$$X_C = \frac{1}{2\pi f C}$$

$$or \ C = \frac{1}{2\pi f X_c} = \frac{1}{2\pi \times 60 \times 3.71} = 715 \times 10^{-6} \ F$$

$$C = 715 \ \mu f \text{ per phase}$$

The capacitor will have to be rated at $460/\sqrt{3} = 266$ V.

(c) For a delta-connected capacitor bank

$$Q_{phase} = Q_{Cph} = \frac{V_{ph}^2}{X_{Cphase}} = \frac{Q_C}{3}$$

$$X_{Cphase} = \frac{V_{ph}^2}{Q_{Cph}} = \frac{V_L^2}{Q_C/3} = \frac{3V_L^2}{Q_C}$$

$$= \frac{3(460)^2}{57000} = 11.14 \ \Omega$$

$$C = \frac{1}{2\pi f X_c} = 238 \ \mu f \text{ per phase}$$

The capacitor must be rated at 460 V.

3.2 POWER MEASUREMENT

The wattmeter is an analog computer that indicates the average power by sensing the voltage, current, and phase relation between the given voltage and current. The wattmeter has two coils, one with many turns of fine wire for a voltage coil, the other with a few turns of heavy wire for a current coil. An indicating needle is attached to one of the coils, and the average magnetic torque on this coil is proportional to the average power between the sensed voltage and current. The magnetic torque is balanced by a spring torque making spring (and needle) deflection proportional to power. The current is connected through the low impedance stationary coils (also called field coils). The voltage is connected across the high impedance movable coil. The pointer attached to the rotating coil displays the interaction between the magnetic fields produced by the currents flowing in the two coils. The instantaneous torque developed is proportional to the instantaneous power, so that

$$p(t) = v(t) \, i(t) \tag{3.12}$$

The following precautions are necessary in using a wattmeter. First, it is possible to exceed the rating of either or both of the coils without exceeding the limit of the wattmeter scale. For example, if one were to use a 1500 W, 150 V, 10 A wattmeter to measure the power into a 240 V, 50 A

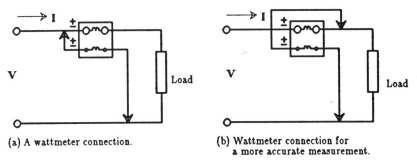

(a) A wattmeter connection.   (b) Wattmeter connection for
                                    a more accurate measurement.

Figure 3.12 Wattmeter connection for upscale indication.

capacitor, the wattmeter would indicate approximately 0 W, yet the voltage coil would be exposed to a voltage 1.6 times its rating and the current coil would be exposed to a current 5 times its rating. This would definitely result in a damaged wattmeter. A second precaution is with regard to the use of wattmeters in low power factor circuits for circuits with low wattage ranges. The voltage coil may have a lower than usual impedance as compared with the measured circuit, and the current coil may have a higher than normal impedance as compared with the measured circuit so that even though a wattmeter may be calibrated for 1 percent accuracy, the reading accuracy may be in error by as much as 20 percent.

Some care must be taken when using a wattmeter since the meter will indicate downscale (or negative) when the angle between the voltage and current is greater than 90°. The modern wattmeter has ± marked on one terminal of each of the voltage and current coils. If the wattmeter is connected as shown in Figure 3.12 for a single-phase load, the indication will be positive or upscale.

A. Measurement of Three-phase Power

To measure power delivered to a three-phase four-wire load, connect three wattmeters as shown in Figure 3.13. The total power is the sum of the three wattmeter indications since each wattmeter measures the phase current and phase voltage of each respective phase. In instantaneous equation form, this total power is

$$p_t = v_{an} i_{Aa} + v_{bn} i_{Bb} + v_{cn} i_{Cc} \qquad (3.13)$$

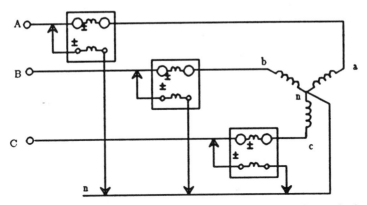

Figure 3.13 Three wattmeters with neutral connection for a 4-wire wye load.

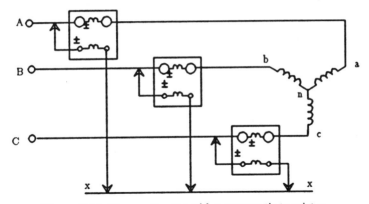

Figure 3.14 Three wattmeters with a common tie to point x.

Now disconnect the wire between n and x to arrive at the circuit of Figure 3.14. The sum of the three wattmeter indications in this circuit is

$$p_x = v_{ax}i_{Aa} + v_{bx}i_{Bb} + v_{cx}i_{Cc} \qquad (3.14)$$

However, the significance of this sum must be questioned. Note that

$$v_{ax} = v_{an} + v_{nx}, \quad v_{bx} = v_{bn} + v_{nx}, \quad \text{and } v_{cx} = v_{cn} + v_{nx}$$

Then (3.14) becomes

$$p_x = \left(v_{an} + v_{nx}\right)i_{Aa} + \left(v_{bn} + v_{nx}\right)i_{Bb} + \left(v_{cn} + v_{nx}\right)i_{Cc} \qquad (3.15)$$

The terms of (3.15) can be rearranged to

$$p_x = v_{an}i_{Aa} + v_{bn}i_{Bb} + v_{cn}i_{Cc} + v_{nx}(i_{Aa} + i_{Bb} + i_{Cc}) \quad (3.16)$$

Now by Kirchhoff's current law at node n, $i_{Aa} + i_{Bb} + i_{Cc} = 0$, which means that the last terms of (3.16) is zero, leaving

$$p_x = v_{an}i_{Aa} + v_{bn}i_{Bb} + v_{cn}i_{Cc} \quad (3.17)$$

which is identical to (3.14). Thus, the wattmeters connected as shown in Figure 3.14 indicate the total power of a three-phase, three-wire circuit. Since the derivation was for instantaneous values, the measurement applies to all circuits whether they be balanced, unbalanced, single frequency, multiple frequency, etc. A second observation is that x is any point in the system. For example, if one connects point x to point b, wattmeter P_2 will always indicate zero and can be removed, leaving the other two wattmeters to indicate total power. Obviously using two wattmeters means less equipment and fewer connections in making measurements. Also, the load circuit may not have the n terminal available as in a delta circuit with only the a, b, and c terminals available. Since all delta circuits can be replaced with an equivalent wye circuit, the two wattmeter method will indicate the total power to any three-wire three-phase system.

The point x of Figure 3.14 can also be tied to point a or point c, thus permitting two other ways of connecting two wattmeters for indicating the total power. A final caution: one must be careful how to connect these two wattmeters in order to get the proper indication of total power. Three methods for connecting two wattmeters for measuring three-phase power are shown in Figure 3.15.

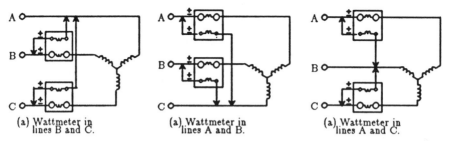

(a) Wattmeter in lines B and C.   (a) Wattmeter in lines A and B.   (a) Wattmeter in lines A and C.

Figure 3.15 Three different connections for the two wattmeter method of measuring three-phase power for a three-wire system.

In general, $(n - 1)$ wattmeters are required for measuring power to an n-wire system. Thus, for a single-phase circuit, we need one wattmeter, for a three-wire three-phase system we need two wattmeters, for a four-wire three-phase system we need three wattmeters, and for a seven-wire six-phase system we need six wattmeters.

Example 3.9

Show the connections for measuring three-phase power for a three-wire three-phase circuit using two wattmeters.

Solution

Begin by connecting two wattmeters for an assumed power flow from left to right as shown in Figure 3.16a. (See Figure 3.12 for description of the individual wattmeter connections.)

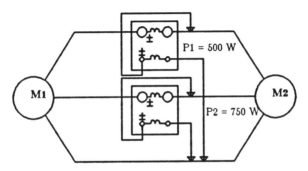

(a) Power flow is to the right since both meters are connected properly and P1 + P2 > 0.

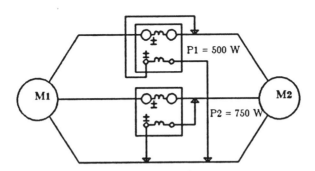

(b) Power flow is to the left since P1 is connected properly with P2 reversed and P2 > P1.

Figure 3.16 Two wattmeter connections for determining power flow for Example 3.9.

Case 1. The meters are connected correctly for power flow from left to right, meter P1 indicates up scale 500 W and meter P2 indicates up scale 750 W with the meters connected as shown in Figure 3.16. Thus the total power is

$$P1 + P2 = 500 + 750 = 1250 \text{ W from left to right}$$

with Machine 1 as the generator.

Case 2. If the load is modified and the meters are left connected as shown in Figure 3.16a, meter P1 will indicate 500 W upscale and meter P2 will indicate an unknown value downscale. In order to obtain a meaningful measurement, change the meter P2 voltage leads as shown in Figure 3.16b. If meter P2 indicates upscale with this connection, this implies a negative quantity for meter P2. With the voltage leads interchanged as shown in Figure 3.16b, wattmeter P1 indicates an upscale value of 500 W (+500) and P2 indicates an upscale value of 750 W. Since the P2 voltage leads are reversed, the value is (-750). Thus,

$$P1 + P2 = 500 + (-750) = -250 \text{ W}$$

This means the total power flow is 250 W from right to left with Machine 2 as the generator.

3.3 ELECTRICAL SYSTEM DESIGN

An engineer should be capable of specifying the appropriate size of conductor, conduit, and circuit protection for a given electrical system. A recognized source of information is the National Electrical Code® (NEC)®. [2]

A. Circuit Breakers

A circuit breaker is an electromechanical device for interrupting an electrical circuit when the electrical current is greater than a designated value. Several molded-case circuit breakers, 60 A frame size, are shown in Figure

---

[2]A few selected tables from the NEC® are included as Appendix B. The NEC® is prepared by the National Fire Protection Association (NFPA) as document NFPA-70. It is also accepted as the American National Standard Institute (ANSI) publication C1. These organizations are nongovernmental. Members of the editorial committees are volunteer representatives from engineering, electrical inspection, electrical contracting, equipment manufacturers, OSHA, testing laboratories, etc. National Electrical Code® and NEC® are registered trade marks of the National Fire Protection Association, Inc., Quincy, MA.

Figure 3.17 Molded case circuit breakers, 60 A frame size.
(a) Tandem, (b) one pole, (c) two pole, (d) three pole. (Courtesy of the Square D Company)

3.17. Also, a circuit breaker panel is shown in Figure 11.5. The *instantaneous trip* breaker includes an electromagnet that causes the breaker to trip on a circuit fault of greater than 700 percent of the designated value. There are restrictions on the application of this type of breaker. Per NEC® Section 430-52:

> An instantaneous trip circuit breaker shall be used only if adjustable, if part of a combination controller having motor-running overload and also short-circuit and ground-fault protection in each conductor, and if the combination is especially approved for the purpose. A motor short-circuit protector shall be permitted in lieu of devices listed in Table 430-152 if the motor short-circuit protector is part of a combination controller having both motor overload protection and short-circuit and ground-fault protection in each conductor if it will operate at not more than 1300 percent of full-load motor current and if the combination is especially approved for the purpose.

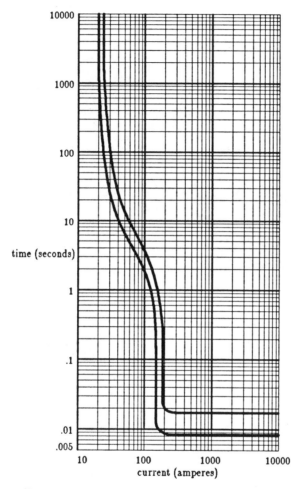

Figure 3.18 Time-current curves for a 20 A molded case circuit breaker.

The *inverse time* breaker has an intentional time delay included with the instantaneous trip feature. This is usually accomplished with a bimetallic assembly that provides a thermal time delay overload. Since circuit breakers are electromechanical devices subject to environmental temperature changes, the tripping time data is given as a band of operation. Each of these breakers has a characteristic curve of clearing time vs. current. Thus the 20-ampere breaker described by the curve of Figure 3.18 would interrupt a circuit carrying 500 amperes within 1/2 to 1 cycle (0.008 to 0.017 seconds). It would take 6 to 15 seconds for 50 amperes to trip the breaker. What is the significance of the 20 ampere rating? A 20-ampere breaker shall interrupt the circuit in less than three hours when a 20 ampere current flows through the breaker. This last statement leads to the NEC® design requirement (Section 220-2a) that the computed continuous branch load (a load connected more than three hours) shall not exceed 80 percent of the branch circuit rating. For a current of ten times that rated, the breaker shall interrupt in less than 1/2 cycle.

B. Fuses

A fuse is a safety device intended to protect (1) the wiring in a building, (2) an electrical apparatus, or (3) an electrical system. An electrical fuse is a metallic link of relatively low melting point that melts when an electric current above a given amount flows through the link. Several styles of fuses are shown in Figure 3.19. Two styles of fuses intended for low voltage use (under 600 V) are (1) instantaneous (or non-time delay) and (2) dual-element (also called time delay or slow-blow). Each of these fuses has a characteristic curve of clearing time vs. current. For example, the 30-ampere fuse described by the first curve of Figure 3.22 would interrupt a circuit carrying 500 amperes in 0.05 seconds (50 ms). It would take 250 seconds (4.2 min) for 50 amperes to melt the fuse.

The *inverse time* fuses (also called *dual element* or *slow blow* fuses) have a time delay as well as an instantaneous feature. On a short-circuit, the fuse link melts just as in the case of the instantaneous types of fuses. On an overload, the fuse link remains entirely inactive. The heat generated by an overload is fed to the center mass of copper on which is mounted a spring and a short connector. This connector is held in place by low melting point solder and connects the center mass of copper to the fuse link. When the temperature is increased by the overload to such an extent as to melt the solder, this connector is pulled out of place, thereby opening the circuit.

C. Branch and Service Conductors

Electrical conductors are sized in the United State according to American Wire Gage (AWG) standards in sizes 4/0, 3/0, 2/0, 1/0, 1... 10, 12, 14, etc. The larger the number, the smaller the conductor. The size 4/0 is spoken as "four-aught" and is also written as "0000" with similar names for the other *aught* sizes. For wire sizes larger than 4/0, the designation is given in circular mils. A circular mil is a circular area with a diameter of 1 mil (1/1000 of an inch). Insulated conductors used for electrical service of sizes

(a) Electronic and automotive fuse.

(b) Plug fuse, 15 A.

(c) Standard 25 A, 250 V cartridge
fuse, 3 inches long.

(d) Rejection type 25 A, 250 V, Class R
cartridge fuse, 3 inches long.

(e) Class R type fuse rejection clips
which accept only the Class R
rejection type fuses.

(f) Current limiting 600 V,
2000 A, 200,000 AIC, 9 inches long.

(g) Knife blade rejection type,
200 A, 250 V, 5¼ inches long.

(h) Power semiconductors used
in solid-state power equipment such
as rectifiers, inverters, dc motor
drives, require very fast,
current-limiting fuse protection.

Figure 3.19 Examples of various styles of fuses. (Reprinted with
permission by Bussmann Division, McGraw-Edison Company)

AWG #10 and smaller are generally solid conductors while AWG #8 and
larger are generally stranded conductors. NEC® Tables 310-16 to 310-19
include standard size conductors with the respective ampacity rating (Table
310-16 is included in Appendix B). For conductors AWG #10 and smaller

(a) Cut-a-way view of typical single-element fuse.

(b) Under sustained overload a section of the link melts and an arc is established.

(c) The "open" single-element fuse after opening a circuit overload.

(d) When subjected to a short-circuit current, several sections of the fuse link melt almost instantly.

(e) The "open" single-element fuse after opening a shorted circuit.

Figure 3.20 Single-element Fuse.

the usual insulation is TW (thermoplastic waterproof). Conductors #8 AWG and larger are normally THW (thermoplastic heat resistant waterproof). The conductor size selected is the smallest size that will handle the load with allowance for voltage drop.

(a) Cut-a-way view of typical dual-element fuse with distinct and separate overload and short-circuit elements.

(b) Under sustained overload conditions, the trigger spring fractures the calibrated fusing alloy and releases the "connector."

(c) Like the single element fuse, a short-circuit current causes the restricted portions of the short-circuit elements to melt and arcing to burn back the resulting gaps until the arcs are suppressed by the arc quenching material and increased arc resistance.

Figure 3.21 Inverse time fuse and regular fuse.

D. Raceway

The generic term "raceway" designates "an enclosed channel designed expressly for holding wires, cables...." Raceways may be of metal or insulated material and include rigid metal conduit, electrical metallic tubing (EMT), and PVC tubing. The maximum number of conductors permitted in conduit is given in NEC® Chapter 9 (see Appendix B).

Example 3.10

A storage building has an 18 kW, three-phase incandescent lighting load. The lights are used more than 3 hours per day. Specify the following according to the NEC®: (a) size of THW copper feeder conductor, (b) size of conduit, (c) size of fuse.

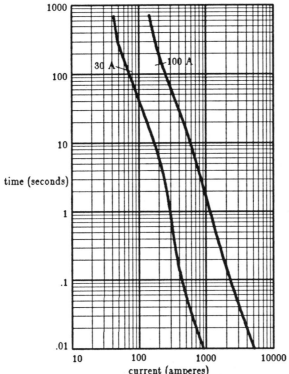

Figure 3.22 Fuse interrupting time curves.

Solution

The incandescent lamp has a power factor of unity. Therefore, the line current is

$$I = \frac{P}{\sqrt{3}\,V\cos\theta} = \frac{18000}{\sqrt{3} \times 208 \times 1} = 50.0 \text{ A}$$

(a) The conductor is sized for 125 percent of full load current (see Appendix B.1), or

$$I = 1.25(50) = 62.5 \text{ A}$$

From NEC® Table 310-16 (Appendix B), use #6 AWG THW copper conductor.

(b) The conduit size for four #6 AWG conductors is 1 inch according to NEC®, Chapter 9, Table 3A (Appendix B).

(c) The fuse is sized to match the wire size at 125 percent of full load or 62.5 A. The code permits using the next larger size, or 70 A fuse according to NEC®, Section 240-6 (Appendix B.3).

Example 3.11

Repeat Example 3.10 for a fluorescent lamp load with a power factor of 95 percent.

Solution

The line current is:

$$I = \frac{18000}{\sqrt{3} \times 208 \times 0.95} = 52.6 \text{ A}$$

(a) The conductor is sized for 125 percent of full load current, or

$$I = 1.25 \times 52.6 = 65.7 \text{ A}$$

From NEC® Table 310-16, Note 10(c), "The neutral shall be considered to be a current-carrying conductor." Then according to Table 310-16, Note 8, 4-6 conductors require a derating of 80 percent. However, according to Note 8, exception 2, both derating factors do not apply. Now according to Table 310-16, a #6 AWG THW conductor would be appropriate.

(b) The conduit for four #6 AWG would be 1 inch according to NEC® Chapter 9, Table 3A.

(c) The fuse is sized for 125 percent of load current, or 65.7 A. Section 240-6 of the NEC® indicates the next largest size of fuse (or circuit breaker) is 70 A. Note that if the load were connected less than three hours at a given time, the 125 percent of full load is not required. However, the requirements of Table 310-16 Note 10(c) and Note 8 do apply and the #6 AWG size conductor would be required.

PROBLEMS

3.1 Given a balanced three-phase wye-connected load with $80\underline{/20°}$ Ω per phase for an *abc* sequence voltage supply with $\mathbf{V}_{ab} = 208\underline{/0°}$ V. (a) What are the magnitudes of the line and phase voltages? (b) What are the magnitudes of the line and phase currents? (c) Sketch a phasor diagram showing all the phase and line voltages and currents. (d) What is the total power required by the load?

3.2 Repeat Problem 3.1 for an *acb* sequence.

3.3 Repeat Problem 3.1 for an *abc* sequence with $V_{bc} = 208\underline{/0^\circ}$ V.

3.4 Repeat Problem 3.1 for an *acb* sequence with $V_{bc} = 208\underline{/0^\circ}$ V.

3.5 Given a balanced three-phase wye-connected load with $80\underline{/70^\circ}$ Ω per phase for an *abc* sequence voltage supply with $\mathbf{V}_{ab} = 208\underline{/0^\circ}$ V. (a) What are the magnitudes of the line and phase voltages? (b) What are the magnitudes of the line and phase currents? (c) Sketch a phasor diagram showing all the phase and line voltages and currents. (d) What is the total power required by the load?

3.6 Repeat Problem 3.5 for an *acb* sequence.

3.7 Given a balanced three-phase delta load with a line voltage of 208 V, *abc* sequence, and an impedance of $80\underline{/20^\circ}$ Ω for each phase. Use V_{ab} at 0°. (a) Show all currents and voltages in a phasor diagram. (b) What is the total power supplied to the load?

3.8 Repeat Problem 3.7 for an *acb* sequence.

3.9 A power company has several power plants feeding into a three-phase network of lines. The power plants are supplying power as follows:

Plant A = 250 MVA at unity power factor

Plant B = 400 MVA at 90 percent power factor lagging

Plant C = 100 MVA at 60 percent power factor lagging

(a) What is the total kVA output? (b) What is the system power factor?

3.10 A city is supplied by three-phase transmission lines from two different power plants. The average load is 20,000 kVA at a power factor of 0.707 (inductive). If one power plant supplies 5000 kVA at unity power factor, what does the other supply?

3.11 The following 2300 V, 60 Hz, three-phase loads are connected in parallel.

Load 1 200 kVA at 0.8 lagging power factor

Load 2 50 kW at unity power factor

(a) What kvar rating of capacitors is required to correct the power factor to 0.95? (b) What is the magnitude of the supply current before the capacitors are added? (c) What is the magnitude of the supply current after the capacitors are added?

3.12 A balanced three-phase load of 600 kW, 60 percent inductive power factor is to be supplemented by a 200 kW synchronous motor. (The synchronous motor may be adjusted to operate at leading or at lagging power factor.) What kVA and power factor of the synchronous motor is required to raise the power factor of the system to 0.90?

3.13 (a) A 480 V three-phase three-wire transmission line supplies power to a wye-connected induction motor. The motor delivers 20 hp and operates at an efficiency of 85 percent and a power factor of 80 percent. If the motor is considered a balanced load, what is the line current? (b) Repeat part (a) for a delta-connected motor.

3.14 What is the power dissipated by the load of Figure 3.23?

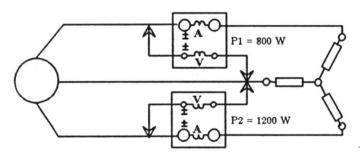

Figure 3.23 Circuit for Problem 3.14.

3.15 What is the power dissipated by the load of Figure 3.24?

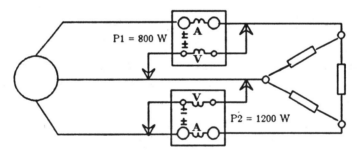

Figure 3.24 Circuit for Problem 3.15.

3.16 In Figure 3.25, which machine is the generator and which machine is the motor? What is the magnitude of the power flow?

3.17 In Figure 3.26, which machine is the generator and which machine is the motor? What is the magnitude of the power flow?

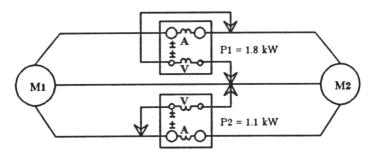

Figure 3.25 Circuit for Problem 3.16.

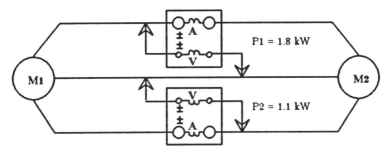

Figure 3.26 Circuit for Problem 3.17.

3.18 In Figure 3.27, which machine is the generator and which machine is the motor? What is the magnitude of the power flow?

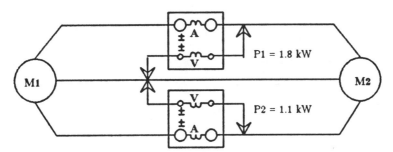

Figure 3.27 Circuit for Problem 3.18.

3.19 In Figure 3.28, which machine is the generator and which machine is the motor? What is the magnitude of the power flow?

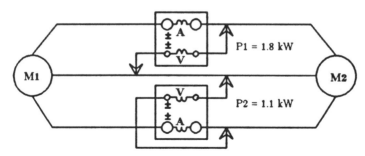

Figure 3.28 Circuit for Problem 3.19.

3.20 Sketch the connections for measuring three-phase power for the circuit of Problem 3.1 with three wattmeters. (a) By referring to the phasors of Problem 3.1 predict the indications of each wattmeter. (b) What is the total power absorbed by the load?

3.21 Show the two-wattmeter connection for measuring the power delivered to the load of Problem 3.1. Connect wattmeter P1 in line *b* and wattmeter P2 in line *c*. (a) What is the wattage indication of each wattmeter? Does the sum of the two wattmeters indicate the total power? (b) Repeat part *a* with wattmeter P1 in line *a* and wattmeter P2 in line *b*. (c) Repeat part *a* with wattmeter P1 in line *a* and wattmeter P2 in line *c*.

3.22 Repeat Problem 3.21 for an *acb* sequence.

3.23 What is the indication of each wattmeter with P1 in line *a* and P2 in line *b* for the load of Problem 3.5? What does the algebraic sum of the two wattmeters represent?

3.24 What is the indication of each wattmeter with P1 in line *a* and P2 in line *b* for the load of Problem 3.6? What does the algebraic sum of the two wattmeters represent?

3.25 What is the wattmeter indication for the circuit of Figure 3.29?

3.26 Given a 460 V, 200 kW, 60 Hz, wye-connected, three-phase, three-wire resistance type electric furnace, (a) determine the magnitude of the line current. (b) What is the resistance (in ohms) of the load per phase?

3.27 Determine the following for Problem 3.26: (a) size of conductor and conduit for THW copper, (b) size of fuse, and (c) size of circuit breaker. (See Appendix B for excerpts of the National Electrical Code® tables.)

3.28 A balanced three-phase wye-connected motor consumes 25 kVA at 80 percent lagging power factor. If the applied line voltage is $V_{ab} = 200 + j100$, solve for the complex current phasor and then plot a

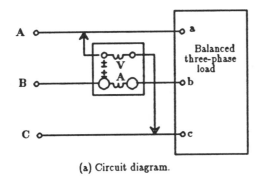

(a) Circuit diagram.

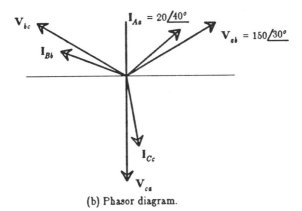

(b) Phasor diagram.

Figure 3.29 Figure for Problem 3.25.

phasor diagram of all voltage and currents. Assume an *abc* sequence.

3.29 Two balanced three-phase delta-connected loads are supplied in parallel from the same feeder. The first load is 40 kVA at 0.90 lagging power factor and the second load is 80 kVA at 0.60 lagging power factor. Assume *abc* sequence. (a) Find P, Q, and S supplied by the feeder. (b) Find the power factor of the combined load. (c) If the applied voltage is $V_{ab} = 2300 + j0$, find the feeder current phasor. (d) Repeat (c) for $V_{ab} = 2300\underline{/60°}$ V.

3.30 A balanced three-phase load of 1000 kVA has a power factor of 0.60 lagging. The system power factor is to be raised to 0.8 by connecting suitable capacitors in parallel. Find: (a) S of the capacitors (capacitor power factor is considered to be zero). (b) Capacitance of delta-connected capacitors for a frequency of 60 Hz and a line voltage of 2300 V. (c) Capacitance of delta-connected capacitors for a frequency of 60 Hz and a line voltage of 115 V. (d) Repeat steps (b) and (c) for wye-connected capacitors.

3.31 Two three-phase generators operating in parallel are delivering power at 240 V. The first generator supplies a current of 18.7 A at 0.85 lagging power factor. The second generator supplies a current of 55.6 A with a 32° leading power factor angle. (a) What percent of the total power is each machine delivering? (b) What is the power factor of the total load? Draw the phasor diagram and the power diagram.

3.32 A three-phase balanced wye-connected load with $(6.31 + j2.51)$ Ω per phase is connected to a 208 V, three-phase, 60 Hz source. Determine (a) the magnitude of the line current and (b) the power dissipated in the load.

3.33 How long does it take for the 30 A fuse of Figure 3.22 to interrupt for a current of (a) 3000 A, (b) 300 A, (c) 30 A?

3.34 Determine the interrupting time for the 100 A fuse of Figure 3.22 for current levels of (a) 2000 A, (b) 200 A, (c) 100 A.

3.35 Determine the tripping time for the 20 A circuit breaker of Figure 3.18 for ampacity levels of (a) 3000 A, (b) 100 A, and (c) 20 A.

3.36 What is the interrupting time for the 30 A fuse of Problem 3.33 and the 20 A circuit breaker of Problem 3.35 connected in series with a load current of (a) 150 A, (b) 200 A, (c) 300 A, (d) 400 A?

3.37 A balanced three-wire delta load with a 5 Ω resistor per phase is connected to a 480 V, three-phase, 60 Hz source. Determine (a) the line current and (b) the power dissipated in the load.

3.38 Determine the following for Problem 3.37: (a) size of copper feeder conductor with THW insulation, (b) size of conduit, and (c) size of molded-case circuit breaker.

3.39 A balanced four-wire wye load with a 5 Ω resistor per phase is connected to a 480 V, three-phase, 60 Hz source. Determine (a) the magnitude of the line current and (b) the power dissipated in the load.

3.40 Determine the following for Problem 3.39: (a) size of aluminum conductor with THW insulation, (b) size of conduit, (c) size of fuse.

3.41 Sketch a circuit for supplying electrical energy to a 115 V, single-phase, incandescent light fixture (a) controlled from two separate switch locations, and (b) controlled from three separate switch locations.

4

POLYPHASE
SYNCHRONOUS
MACHINES

A simple single-phase synchronous machine similar to the emergency power supply generators available from many mail-order houses consists of a heavy winding in a stationary assembly called a *stator* with an electromagnet mounted as part of a rotational assembly called the *rotor*. As the rotor is moved, a current is induced in the stator winding. This induced stator current varies in magnitude and direction with time, hence the name *alternating current*. If additional stator coils are included, the machine is said to be a polyphase machine; the most common type is the three-phase machine.

The synchronous generator has been known by other names such as the *alternator*. Some generators are constructed so as to operate from a slowly driven vertically mounted hydraulic (hydro) water turbine and are called *hydroelectric generators* or *hydro-alternators* (Figure 4.1). Generators driven by fossil- or nuclear-powered steam turbines (Figures 4.2, 4.3, 4.4, 4.5 and 4.6) or fossil-fueled "gas turbines" (Figures 4.7 and 4.8) operating at 1800 rev/min or 3600 rev/min are called *turbo-alternators* or *turbo-generators*. The IEEE recommended name is *synchronous generator*.

If a balanced polyphase voltage is supplied to the stator windings and the synchronous machine is brought up to speed, as will be described later, the machine will rotate as a motor at a speed proportional to the frequency of the supply voltage. If the supply frequency is held constant the machine will rotate at a constant or synchronous speed, hence the name *synchronous motor* (Figure 4.9).

Figure 4.1 Hydroelectric synchronous generator with rotor removed for inspection. *(Courtesy of General Electric Company)*

Figure 4.2 Large steam turbine synchronous generator. *(Courtesy of General Electric Company)*

A synchronous motor operated at zero leading power factor and constructed without an external shaft is called a *synchronous condenser* or a *synchronous capacitor* and is used by utility power companies for system stability and power factor correction.

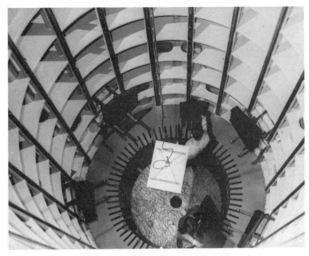

Figure 4.3 Installing the generator laminations in a large steam turbine synchronous generator. *(Courtesy of General Electric Company)*

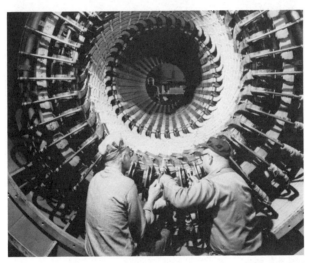

Figure 4.4 Completing the winding of a large steam turbine synchronous generator. *(Courtesy of General Electric Company)*

4.1 INTRODUCTION TO SYNCHRONOUS GENERATORS

All rotating electric generators operate on the principle of induced voltage described by (1.1) and rewritten here as (4.1).

Figure 4.5 Rotor for a large steam turbine synchronous generator rated 724 MVA, 3600 rev/min, 24 kV. Generator rotors must have a mechanical balance within very close tolerances to prevent excessive vibration, even though they weigh over 165 tons and measure 45 feet long. This rotor is being moved into position for test. *(Courtesy of General Electric Company)*

Figure 4.6 Machining the conductor slots in the rotor of a large steam turbine synchronous generator. *(Courtesy of the General Electric Co.)*

$$\vec{e} = l \left(\vec{u} \times \vec{B} \right) \tag{4.1}$$

For the synchronous generator (4.1) can be rewritten as

$$E_{af\ (rms)} = \frac{2\pi}{\sqrt{2}} fNK_W \phi_{\max} = 4.44fNK_W \phi_{\max} \tag{4.2}$$

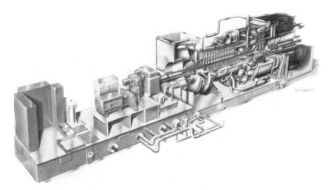

Figure 4.7 Combustion gas turbine, simple-cycle, single-shaft, 50 MW size. *(Courtesy of General Electric Company)*

Figure 4.8 Floating power plants. Barge-mounted gas turbines featuring maximum silencing, appearance-designed housing and fuel systems that emit no visible exhaust, offer flexibility in location and permit easy removal when required. In addition, the prepackaging, pretesting and fuel flexibility of the barge-mounted, heavy-duty gas turbine make it increasingly attractive for utility applications. *(Courtesy of General Electric Company)*

where N is the number of series turns per phase, f is the frequency of the supply voltage in hertz, ϕ_{max} is the maximum magnetic flux in webers, and K_W is the winding factor for distributed windings (between 0.85 and 0.95 for most three-phase windings). The winding factor includes distribution, pitch, and skew factors that are not described in this text. These are design features that are not necessary for the application of a synchronous

Figure 4.9 Synchronous motor. *(Courtesy of General Electric Company)*

machine and will be left for discussion in texts specializing in such details
[A14C].

A. Structural Features

The field (or rotor) of a synchronous machine may be of (a) salient or (b)
nonsalient construction as indicated in Figure 4.10. One definitions for the
word salient is that which projects outward or upward from its surround-
ings. Generators with more than four poles are usually of the *salient* type
(see Figs 4.1 and 4.9). Steam turbine generators are generally of the *nonsalient* type (see Figs 4.2 to 4.7). The nonsalient pole rotors are also called
round rotors, or *cylindrical* rotors.

The direct current required for the field may be supplied from (1) a
dc generator (called an exciter), which is generally connected to the shaft of
the synchronous machine, (2) an alternator (with rectifiers) connected to
the shaft of the synchronous generator, and (3) rectifying a portion of the
generator output.

The field energy is supplied through *slip rings* (also called *collector
rings)* to the field in most machines. However, a synchronous generator
exciter may be constructed with the exciter line conductors and rectifiers
on the rotor and the exciter field as the stationary part. The efficiency of
the large steam turbine generators can be increased by allowing the field to

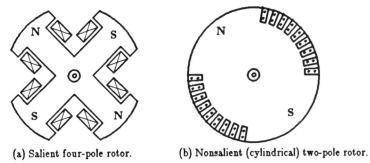

(a) Salient four-pole rotor.   (b) Nonsalient (cylindrical) two-pole rotor.

Figure 4.10 Synchronous machine salient-pole and nonsalient-pole rotors.

rotate in a hydrogen atmosphere. Although hydrogen in concentrations of 30 to 70 percent in air is hazardous, the nearly 100 percent concentration of hydrogen has been used since the 1930s with very few incidents. This feature is referred to as *hydrogen cooling.*

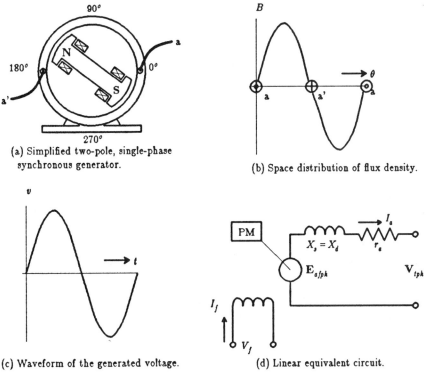

(a) Simplified two-pole, single-phase
synchronous generator.

(b) Space distribution of flux density.

(c) Waveform of the generated voltage.   (d) Linear equivalent circuit.

Figure 4.11 Single-phase, two-pole synchronous machine.

The stator windings that carry the synchronous generator output current are embedded in slots in the stator assembly. This assembly consists of a laminated core (Figures 4.3 and 4.4) supported in a frame. On large machines the use of water circulated through the stator conductors has allowed a five-fold increase in generator capacity for the same size structure.

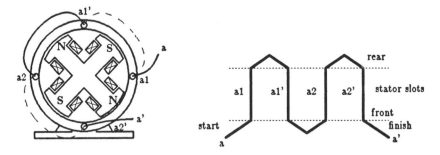

(a) Simplified salient four-pole, single-phase, synchronous machine.

(b) Developed view of stator windings.

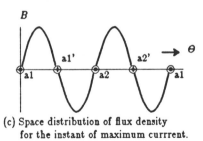

(c) Space distribution of flux density for the instant of maximum currrent.

Figure 4.12 Simplified single-phase, four-pole synchronous machine.

The simple single-phase machine with one coil on the stator and two poles for the rotor was shown in Figure 4.11. A four-pole single-phase machine with one coil per pair of poles is shown in Figure 4.12. A developed view of the two coils (four coil sides) is shown in Figure 4.12b. The magnetic flux density distribution for the instant of maximum current flow is shown in Figure 4.12c. Note that there are 720 electrical degrees (4π radians) for 360 mechanical degrees (2π radians).

The majority of machines have several coils per phase in a form called distributed windings. These windings are found in spiral, lap, and wave shown in Figure 4.13. The spiral type is also called the concentric winding and is used extensively in small machines. The coils may have several turns per coil to increase the voltage per total coil and are shown in Figure 4.14.

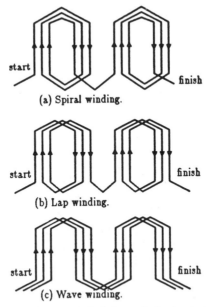

(a) Spiral winding.

(b) Lap winding.

(c) Wave winding.

Figure 4.13 Elementary single-layer single-phase windings.

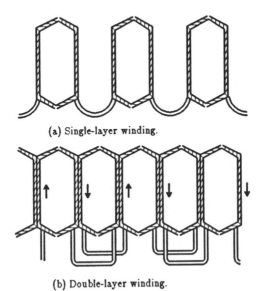

(a) Single-layer winding.

(b) Double-layer winding.

Figure 4.14 Typical synchronous machine stator windings.

A three-phase machine has three sets of coils per pair of poles. The two pole machines of Figure 4.15a has three coils (six coil sides) for one conductor per pole per phase. A stator with a distributed winding coil has more conductors per phase. A four-pole generator with one conductor per pole per phase as shown in Figure 4.15b show two coils per phase, six coils for the three phases, or 12 coil sides. The two coils per phase connected in series form a wye-connected configuration of the four pole generator as shown in Figure 4.15c.

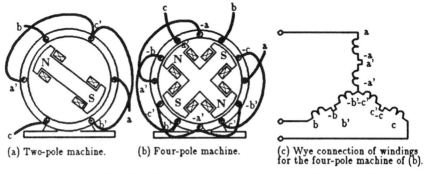

(a) Two-pole machine.   (b) Four-pole machine.   (c) Wye connection of windings for the four-pole machine of (b).

Figure 4.15 Simple salient, three-phase, one coil per phase, synchronous machine.

B. Generator Action

The frequency of a generator output is a function of the rotor speed and the number of poles. As an equation this is

$$f = \frac{p}{2} \times \frac{n}{60} \tag{4.3}$$

where f = frequency in hertz, p is the number of poles, and n is the speed in revolutions per minute.

C. Linear Mathematical Model

A simple mathematical model of a synchronous machine is the Thevenin's equivalent shown in Figure 4.16. The following example outlines the quantitative feature of a synchronous machine used as a motor or as a generator. The resistance value has been exaggerated so that all phasors can be readily observed.

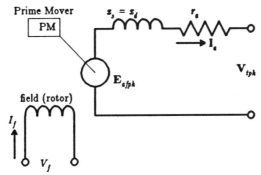

Figure 4.16 Linear single-phase equivalent circuit for a three-phase synchronous machine.

Example 4.1

A 100 kVA, 2300 V, 600 rev/min, three-phase, wye-connected synchronous machine has an effective stator resistance of 4 Ω and a synchronous reactance of 11 Ω and is operating as an isolated generator not part of a large power system. Determine the excitation voltage E_{af} for rated line current at rated terminal voltage when the machine is operating (a) as a generator at unity power factor, (b) as a generator at 80 percent lagging power factor, and (c) as a generator at 60 percent leading power factor.

Solution (See Figure 4.17).

The rated line current for the machine is:

$$I_a = \frac{100 \text{ kVA}}{\sqrt{3} \times 2.3 \text{ kV}} = 25.10 \text{ A}$$

The rated terminal voltage (line-to-neutral) is $2300/\sqrt{3} = 1328$ V.

(a) The excitation voltage for the machine operating as a generator at unity power factor is:

$$\mathbf{E}_{afph} = \mathbf{V}_{tph} + \mathbf{I}_a \left(r_a + jX_s \right)$$

$$= 1328\underline{/0°} + (25.1\underline{/0°})(4 + j11)$$

$$= (1328 + j0) + (100.4 + j276.1)$$

$$= 1455\underline{/10.94°} \text{ V line-to-neutral}$$

$$E_{af} = \sqrt{3} \times 1455 = 2520 \text{ V line-to-line}$$

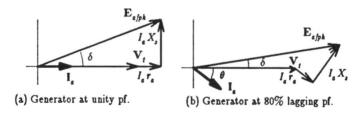

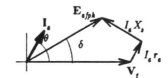

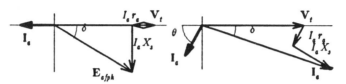

(c) Generator at 60% leading pf.

(d) Motor at unity pf.   (e) Motor at 60% leading pf.

Figure 4.17 Phasor diagram for Examples 4.1 and 4.2.

(b) The excitation voltage for the machine operating as a generator at 80 percent lagging power factor is:

$$\mathbf{E}_{af\ ph} = \mathbf{V}_t + \mathbf{I}_a\left(r_a + jX_s\right)$$

$$= 1328\underline{/0^\circ} + (25.10\underline{/-36.87^\circ})(4 + j11)$$

$$= 1582\underline{/5.83^\circ} \text{ V line-to-neutral}$$

$$E_{af} = \sqrt{3} \times 1582 = 2740 \text{ V line-to-line}$$

(c) The excitation voltage for the machine operating as a generator at 60 percent leading power factor is:

$$E_{afph} = V_{tph} + I_a\left(r_a + jX_s\right) = 1328\underline{/0^\circ} + (25.10\underline{/53.13^\circ})(4 + j11)$$

$$= 1193\underline{/11.90^\circ} \text{ V line-to-neutral}$$

$$E_{af} = \sqrt{3} \times 1193 = 2066 \text{ V line-to-line}$$

4.2 ROTATING MAGNETIC FIELDS

The basis of operation of all polyphase alternating current motors is the rotating magnetic field. The interpretation of the rotating magnetic field requires a multidimensional presentation. This will be done by making two plots. The first is a plot of current vs. time and the second a plot of magnetic flux density vs. angular position of the machine stator. Begin by considering a single phase coil as shown in Figure 4.18a. Assume at time t_3 the current is entering at a and leaving at a'. The magnetic flux density will be down between a and a' and up between a' and a as shown in Figure 4.18d. In the practical machine there are several coils. The flux density patterns similar to Figure 4.18d will add to produce a sinusoidal distribution as shown in Figure 4.18e. The sinusoidal shape will be used for the remaining explanations. Figure 4.18c is an attempt to show the flux density with respect to the stator circumference. Note that the flux density at time t_3 will be out between a and a' and in between a' and a as shown in Figure 4.18e. As the current magnitude varies with time, the height of the flux density curve varies as shown in Figure 4.18f. For example: at time t_0 there is no flux density, at time t_1 the flux density will be 0.5 maximum, at time t_2 the flux density will be 0.87 maximum, and at time t_3 the flux density will be maximum.

Now consider stator windings for the simplified three-phase two-pole machines with windings 120° apart as shown in Figure 4.19. The instantaneous flux density patterns for several time instances are shown in 4.19c, d, and e. Consider, for example, the flux density at time t_0. Note that the current in phase a is zero, the current in phase b is a minus 0.866 maximum, and the current in phase c is plus 0.866 maximum. The related flux density patterns are shown in Figure 4.19c. The resultant flux density of the three flux densities is the heavy line of the figure and is a sine wave of the same period but of different magnitude and phase. Repeat this procedure for the other instants of time and then observe the nature of the resultant waves. It will be noted that the flux density is of constant magnitude but has moved from the left to right on the page (Figure 4.19f). This same thing could be accomplished by moving the constant flux density around the stator. Hence, the three coils with balanced three-phase currents will produce a rotating magnetic field about the stator. A two-phase system is another common system for producing a rotating magnetic field. In fact, any balanced polyphase system will produce the same effect.

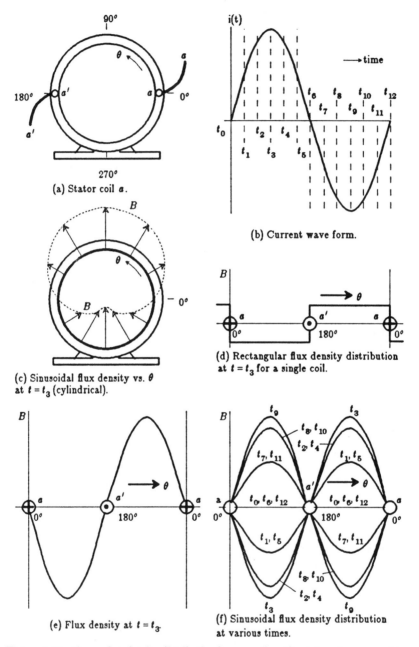

(a) Stator coil *a*.

(b) Current wave form.

(c) Sinusoidal flux density vs. *θ* at *t* = *t*$_3$ (cylindrical).

(d) Rectangular flux density distribution at *t* = *t*$_3$ for a single coil.

(e) Flux density at *t* = *t*$_3$.

(f) Sinusoidal flux density distribution at various times.

Figure 4.18 Air gap flux density distribution for one coil on the stator of an ac machine.

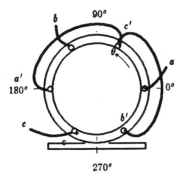

(a) Simplified three-phase
two-pole stator winding.

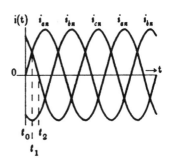

(b) Instantaneous phase currents.

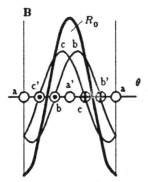

(c) Flux density distribution at time t_0

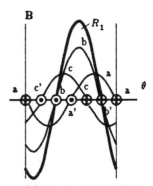

(d) Flux density distribution at time t_1.

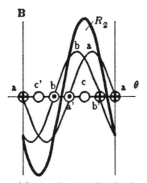

(e) Flux density distribution at time t_2

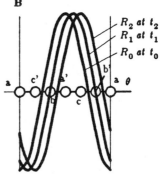

(f) Resultant flux density distribution
at t_0, t_1, and t_2

Figure 4.10 Stator component and resultant field distributions
showing the principle of the rotating field for a three-phase machine.

The speed at which the field rotates is the same as the supply frequency for the given two pole machines. Thus for a 60 Hz supply the field will rotate at 60 rev/s or 3600 rev/min. The machine can be constructed to have more than one set of coils. For each set of coils there is an additional set of poles. Thus the speed is determined by using (4.3) and solving for speed as per (4.4)

$$n = \frac{120f}{p} \tag{4.4}$$

where n is rev/min, f is frequency in Hz, and p is the number of poles.

4.3 SYNCHRONOUS MOTORS

Any synchronous generator may be operated as a motor. The two major advantages for this type of motor are its adjustable power factor, and constant speed for a constant supply frequency. A simple mechanical analogy may prove helpful in explaining how this type of motor operates. The mechanical ring shown in Figure 4.20(a) is caused to rotate at 3600 rev/min counterclockwise. The steady state speed of the rotor that is connected to the outside ring with springs is 3600 rev/min. If a torque is applied to the rotor shaft, the rotor will be displaced momentarily, possibly with some oscillation. After the oscillation has damped out, the rotor will assume a steady state angular position behind the reference axis. The angle of the position is referred to as the torque angle, δ. Consider this analogy further. If the torque is increased sufficiently the mechanical springs will break and the rotor will then stop. This process could be called losing synchronism, and the torque at which this occurs is called *pull-out torque*.

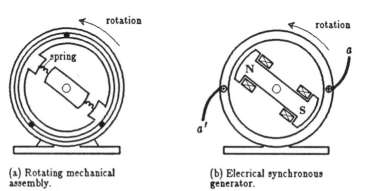

(a) Rotating mechanical assembly.

(b) Elecrical synchronous generator.

Figure 4.20 Mechanical analogy for a synchronous motor.

Now consider the synchronous machine as a motor. The stator with a rotating field is similar to the mechanical ring. Also, the dc electromagnet rotor of the motor is similar to the mechanical rotor with the springs. With no torque connected to the rotor shaft the rotor will move in synchronization with the stator rotating field. When a torque is applied to the shaft, the electromagnet will lag behind by a torque angle, δ. Note that if the torque angle is increased beyond 90°, the magnetic attraction is south-to-south and north-to-north and the machine will pull out of synchronism. This is shown as the sinusoidal curve on Figure 4.21, Curve A. For the salient pole machine there is a reluctance torque that causes the resultant torque shown on Figure 4.21, Curve B. This *pull-out torque* is the maximum steady state torque that the synchronous motor can produce without losing synchronism.

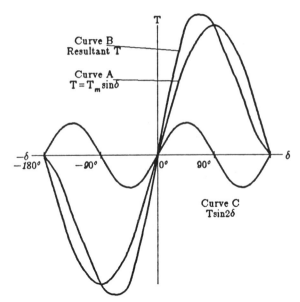

Figure 4.21 Torque angle characteristic of a salient-pole synchronous machine showing fundamental component due to field excitation and second-harmonic component due to reluctance torque.

A synchronous motor, *per se*, has no starting torque. Thus, some external influence is required. Most synchronous motors have a damping (or amortisseur) winding. This winding is similar to a squirrel-cage superimposed about the rotor. This winding allows the machine to be started as a polyphase induction motor as explained in Chapter 7. When the rotor

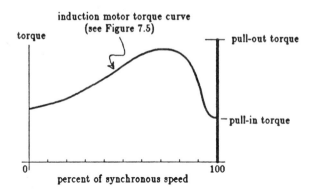

Figure 4.22 Torque-speed characteristic for a synchronous motor with amortisseur windings.

reaches a speed close to a synchronism there is adequate torque to cause the rotor to reach a synchronous speed. The magnitude of this torque is called the *pull-in torque* as indicated in Figure 4.22.

There are three methods for starting a synchronous motor. The first is to start the machine as an induction motor. The procedure for starting in this manner is as follows:

1. Make certain that the field does not develop excessive induced voltage by placing a resistance across the field terminals or shorting the terminals if the induced current is not excessive.

2. When the motor has reached a maximum speed near synchronous, a dc supply is connected to the field while the shorting impedance is removed.

3. The field current is then varied for the desired operating power factor.

A second method of starting a synchronous motor when the machine does not have the amortisseur winding is to have the machine brought up to speed by some type of prime mover connected to the shaft. When the speed, voltage, etc. are properly adjusted, the machine lines are connected to the line in the same manner that they would be connected to any power system as a generator. The mechanical driving torque is then removed and the machine operates as a motor.

A third method is to electrically connect an isolated turbine-driven synchronous generator with the synchronous motor. The turbine, generator, and motor are then brought up to synchronous speed as a system and the synchronous generator eventually is synchronized with the power grid.

The power factor of the synchronous motor may be adjusted by changing the magnitude of the dc field excitation. The following examples introduce the voltage-current phasors related to this idea.

Example 4.2

Repeat Example 4.1 for the synchronous machine operating as a motor and drawing rated line current (a) at unity power factor, and (b) at 60 percent leading power factor. Note that the positive direction of the current is out of the machine at terminal a as shown in Figure 4.16. This procedure aids the appropriated use of mmf phasors.

Solution (See Figures 4.17(d) and 4.17(e))

(a) The excitation voltage for the machine operating as a motor at unity power factor is

$$\mathbf{E}_{afph} = \mathbf{V}_{tph} + \mathbf{I}_a \left(r_a + jx_s \right)$$
$$= 1328\underline{/0^\circ} + (-25.10\underline{/0^\circ})(4 + j11)$$
$$= 1258\underline{/-12.68^\circ} \text{ V line-to-neutral}$$
$$E_{af} = \sqrt{3} \times 1258 = 2178 \text{ V line-to-line}$$

(b) The excitation voltage for the machine operating as a motor at 60 percent leading power factor is:

$$\mathbf{E}_{afph} = \mathbf{V}_{tph} + \mathbf{I}_a \left(r_a + jx_s \, r \right)$$
$$= 1328\underline{/0^\circ} + (-25.10\underline{/53.13^\circ})(4 + j11)$$
$$= 1509\underline{/-9.38^\circ} \text{ V line-to-neutral}$$
$$E_{af} = \sqrt{3} \times 1509 = 2613 \text{ line-to-line}$$

4.4 PARALLEL OPERATION OF SYNCHRONOUS GENERATORS

The typical voltage-load curve for a single-phase synchronous generator connected to a passive load with the speed held constant is shown in Figure 4.23. Consider that there are two controls: the prime move speed control (throttle) and the synchronous generator dc field control. The speed control will cause a change in frequency and voltage magnitude while the generator field control will cause a change in voltage magnitude. The power factor angle is determined by the type of load, whether it be inductive, capacitive, and/or resistive (see Problems **4.11** through **4.13**). When many generators are connected in parallel, the results are much different.

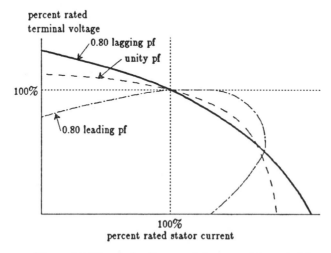

Figure 4.23 Terminal voltage vs. stator current for an isolated synchronous generator with a passive load. The generator is operated at constant speed and with fixed dc excitation.

The electric power systems throughout the world consist of many synchronous generators in parallel. The procedure for connecting a generator with a large system requires only a few basic considerations. The generator and the system must have the same frequency, phase sequence, phase angle, and same terminal voltage. A large generator is connected to a system by automatic methods. For manual connections a synchroscope can be used; however, the phase sequence must be determined independently. A system of three lights called the "dark lamp method" as shown in Figure 4.24a is probably the simplest device for determining proper closing time. The theory of operation of this lamp method is explained in Example 4.3.

Example 4.3

Use the *dark lamp* method shown in Figure 4.24a to synchronize two three-phase synchronous generators. Assume each machine is adjusted for a terminal voltage of 208 V. Determine the voltages across each of the

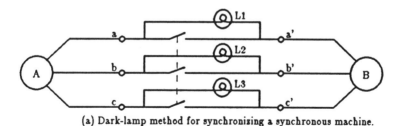

(a) Dark-lamp method for synchronizing a synchronous machine.

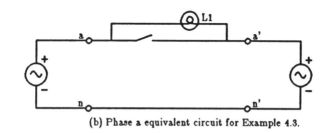

(b) Phase a equivalent circuit for Example 4.3.

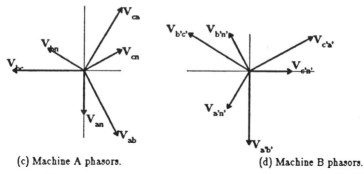

(c) Machine A phasors. (d) Machine B phasors.

Figure 4.24 Dark lamp method for synchronizing an alternator. Machine *A* leads machine *B* by 30°.

synchronizing lamps when the voltages of machine A lead the voltages of machine B by 30°. Let voltage $V_{ab} = 208\underline{/-60^\circ}$ V.

Solution

The phase *a* equivalent circuit for Figure 4.24a is shown in Figure 4.24b. The loop equation is

$$-\mathbf{V}_{an} + \mathbf{V}_{aa'} + \mathbf{V}_{a'n'} = 0$$

or

$$\mathbf{V}_{aa'} = \mathbf{V}_{an} - \mathbf{V}_{a'n'} = 120\underline{/-90^o} - 120\underline{/-120^o}$$

$$= 60 - j16 = 62.12\underline{/-15^o} \text{ the voltage across light 1}$$

$$\mathbf{V}_{bb'} = \mathbf{V}_{bn} - \mathbf{V}_{b'n'} = 62.12\underline{/-135^o} \text{ the voltage across light 2}$$

$$\mathbf{V}_{cc'} = \mathbf{V}_{cn} - \mathbf{V}_{c'n'} = 62.12\underline{/105^o} \text{ the voltage across light 3}$$

Comments. The two machines are in phase when the lamps are dark. When the machines are 180° out of phase the maximum voltage is twice line to neutral or 240 V, therefore the lamps must be special high voltage lamps. If the operator closes a machine into a system for this 180° phase shift the resulting currents have been known to create torques large enough to break the bolts holding the generator to the mounting floor. If the two machines are of opposite phase sequence, then two lights will be bright and one dark. When the two machines are driven at slightly different speeds, these lights will alternate giving the appearance of "circling lights."

It is necessary to measure the voltage of each system as well as to properly synchronize the lights. The appropriate operation also requires the dc field control of the generator and the prime mover throttle control. These functions are better understood following a discussion of the air gap flux density phasors in the following section.

4.5 MAGNETOMOTIVE FORCE RELATIONSHIP

Magnetic flux density in the air gap is the sum of the flux density produced by the dc field and the flux density produced by the stator current. These magnetic components are sinusoidal and are out of phase with each other. Thus the resulting flux density, B_R, is the sum of the field flux density, B_F, and the stator reaction flux density, B_A.

$$B_R = B_F + B_A \tag{4.5}$$

Since the magnetomotive force (ampere-turns) is proportional to the flux density ($B = \mu H$, $B = \mu NI/l$, *or B* is proportional to the mmf) then

$$F_R = F_F + F_A \tag{4.6}$$

The magnetomotive force phasors lead the voltage phasors by 90°. This is derived from the expression

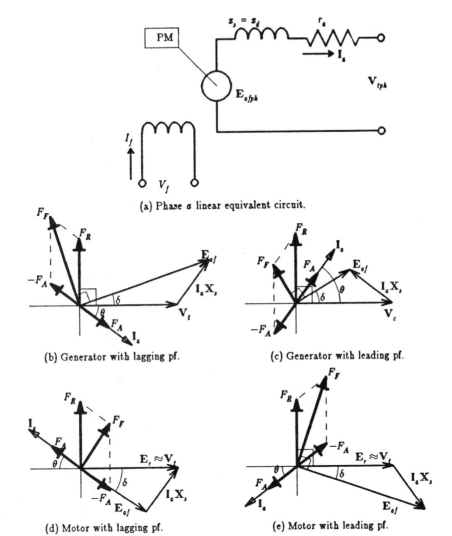

(a) Phase a linear equivalent circuit.

(b) Generator with lagging pf.

(c) Generator with leading pf.

(d) Motor with lagging pf.

(e) Motor with leading pf.

Figure 4.25 Synchronous machine phasor diagrams.

$$v(t) = \frac{d\left[N(t)\phi(t)\right]}{dt} \tag{4.7}$$

or

$$\phi(t) = \frac{1}{N} \int v(t)\,dt$$

If $v(t) = V_m \sin \omega t$, then

$$\phi(t) = \frac{1}{N} \int V_m \sin \omega t \, dt = \frac{-V_m}{\omega N} \cos \omega t$$

$$= \frac{-V_m}{\omega N} \sin(\omega t + 90^\circ)$$

$$= \phi_m \sin(\omega t + 90^\circ) \tag{4.8}$$

If the voltage phasor is V, the flux density phasor is B, then the flux density phasor leads the voltage phasor by 90°.

The diagram showing the voltage, magnetomotive force, and current phasors is shown as Figure 4.25. Begin with the terminal voltage, V_t, and consider the case of a lagging power factor current, I_a. The resultant flux density, B_R, and the magnetomotive force phasor, F_R, lead the terminal voltage by 90°. The stator reaction flux density phasor, B_A, and the magnetomotive force phasor, F_A, are in phase with the stator current phasor I_a. The field equivalent generated flux density B_F and generated magnetomotive force F_F is then the phasor difference between F_R and F_A from (4.6) so that:

$$F_F = F_R - F_A = F_R + (-F_A) \tag{4.9}$$

Now consider that the generated voltage (sometimes called the "voltage behind the impedance") lags behind the generated magnetomotive force F_F by 90°. The length of E_{af} is such that the voltage drop across the reactance, $I_a X_s$, leads I_a by 90°. These phasors are all shown in Figure 4.25a. The phasor diagram has been repeated in Figure 4.25c for the generator operating at leading power factor and again in Figures 4.25d and 4.25e for the motor operating with lagging and leading power factor. These ideas may be applied to the operation of a synchronous machine being operated as a generator supplying energy to a large system or as a motor supplied by the larger system. The details are implied in Problems 4.14 and 4.24 and given in Example 4.4.

Example 4.4

A synchronous generator is supplying energy to a large interconnected system. The field current is adjusted so that the stator current lags the terminal voltage. Neglect stator resistance and leakage reactance.

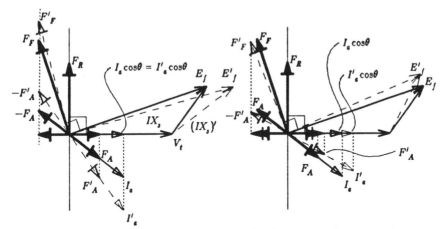

(a) Phasors when the field excitation of the generator is increased 20%.

(b) Phasors when the throttle of the prime mover is increased 20%.

Figure 4.26 Example 4.4.

Figure 4.27 The control of a large interconnected power system is similar to the control of this team of Budweiser Clydesdale horses. *(Courtesy of Fleishman-Hillard, Inc.)*

(a) Increase the field current by 20 percent without changing the driving torque of the prime mover. Qualitatively, what changes occur in power output, in magnitude and phase of the stator current, and in magnitude of the torque angle? Explain by means of phasor diagrams representing the flux density and mmf waves.

(b) Increase the driving torque of the prime mover by 20 percent, without changing the field current. Explain the changes.

Solution

Begin with phasor diagram of Figure 4.25a.

(a) When the machine is connected to a large system, the terminal voltage and system frequency remain constant. This means that the resultant magnetomotive force F_R is a constant. Since the throttle of the prime mover is unchanged and the machine speed is a constant, the power delivered by the generator to the system must remain constant. This means that the phasors $(I_a\cos\theta)$ and $(-F_A\cos\theta)$ are constant. If the dc field current is increased approximately 20 percent, then the value of F_F is increased approximately 20 percent. However, the horizontal projection of F_F which is $(-F_A\cos\theta)$ is a constant. This means that the angle of F_F must be changed so that the end of the F_F phasor must lie on the vertical projection of $-F_A\cos\theta$. This means that end of $-F_{A'}$ also lies on the vertical projection of $-F_A\cos\theta$. Accordingly, the phasor $F_{A'}$ and $I_{a'}$ lie in the opposite direction to $-F_{A'}$ with the end of $I_{a'}$ lying on the vertical projection of $I_a\cos\theta$. Finally, $E_{f'}$ is plotted $90°$ lagging $F_{F'}$ and the length so that $(I_aX_s)'$ is $90°$ lagging of $F_{A'}$.

(b) Begin with the same phasor diagram used at the beginning of part *a*. When the throttle is increased the generator will tend to speed up. However, the system will keep the generator at a constant speed with the result that more power will be delivered from the prime mover through the generator to the electrical system. This means that the phasor $I_a\cos\theta$ will increase in length. This means that $(-F_A\cos\theta)'$ will increase. The phase F_R will remain fixed at $90°$ leading the terminal (and system) voltage. Since the dc field of the generator remains constant, the length of $F_{F'}$ remains constant. Yet the angle of $F_{F'}$ must change so that the end of the $F_{F'}$ phasor lies on the vertical projection of $(-F_A\cos\theta)'$. These changes influence the angle and magnitude of F_A and I_a.

This leads to a conclusion for parts *a* and *b* of this example. If the dc field excitation of the generator is adjusted, the power factor of the generator output is adjusted. If the throttle of the prime mover is adjusted, the power output of the generator is adjusted in magnitude and power factor.

An extension of this idea leads to an understanding of the synchronous motor power factors. If the above procedure is repeated for a sufficient number of values of ac load current and dc field current, a plot

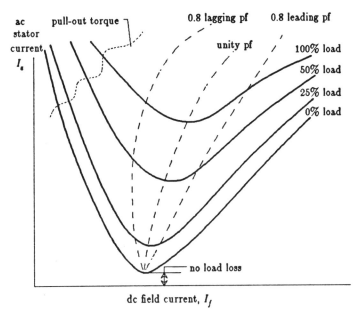

Figure 4.28 Synchronous motor V curves.

called the *V curves* will result with typical data shown in Figure 4.28. To explain the meaning of this curve, consider the case of no mechanical load (the lowest curve of the figure). There is a value of dc field current where the ac stator current is a minimum. This is the unity power factor position. As the dc excitation is decreased the stator current rises and the machine is operating in the lagging power factor mode. If the dc field current is increased from the unity point, the ac stator current will increase and the motor is operating in the leading power factor mode. An easy memory trick is to recall that a synchronous motor without dc excitation appears to the source as a group of coils and is inductive, hence lagging power factor is on the left side of the curve.

PROBLEMS

4.1 Name the major parts of a synchronous machine and describe their functions.

4.2 A hydroelectric power station has several 150 MVA, 125 rev/min generators. How many poles are there on the rotor of each machine for a 50 Hz system?

4.3 What is the maximum speed of a rotating magnetic field in revolutions per minute and in radians per second for a three-phase synchronous machine connected to a (a) 25 Hz supply? (b) for a 60 Hz supply? (c) for a 400 Hz supply?

4.4 A 16,000 kVA, 11 kV, 50 Hz, three-phase, non-salient pole, wye-connected, synchronous generator has a synchronous reactance $X_s = 8.0 \ \Omega$ per phase and stator resistance $r_a = 0.06 \ \Omega$ per phase. (a) Compute the ac excitation voltage, E_{af}, when the generator delivers rated kVA at 0.80 lagging power factor with rated terminal voltage. (b) Repeat part *a* for unity power factor. (c) Repeat part *a* for 0.6 leading power factor.

4.5 A 150 MVA, 20 kV, 125 rev/min, three-phase, wye-connected, synchronous generator is operating at rated voltage, rated stator current, and 0.7 leading power factor. The stator resistance is $r_a = 0.01 \ \Omega$ per phase and the synchronous reactance is $x_s = 2.7 \ \Omega$ per phase. Find the voltage behind the impedance, E_{af}.

4.6 A 150 MVA, 20 kV, 125 rev/min, three-phase, wye-connected, synchronous generator is operating at rated voltage, half of rated current, and 0.7 leading power factor. The stator resistance is $r_a = 0.01 \ \Omega$ per phase and the synchronous reactance is $x_s = 2.7 \ \Omega$ per phase. Find the voltage behind the impedance, E_{af}.

4.7 A nuclear-powered steam turbine synchronous generator is rated 819 MVA, 0.950 power factor, 26 kV, three-phase, 60 Hz, 3600 rev/min, 75 psig. The series impedance is $(0.0010 + j1.0217) \ \Omega$. The generator is supplying rated stator current at rated voltage and 0.84 lagging power factor. (a) What is the excitation voltage? (b) What does the line-to-line terminal voltage become if the load is removed?

4.8 Repeat Problem 4.7 for a load power factor of (a) 0.7 leading and (b) unity.

4.9 A 150 MVA, 20 kV, 125 rev/min, three-phase, synchronous generator operating at rated voltage is supplying a three-phase wye load with an impedance of $(2.0 + j1.2) \ \Omega$ per phase. The stator resistance is $r_a = 0.01 \ \Omega$ per phase and the synchronous reactance is $x_s = 2.7 \ \Omega$. Find the voltage behind the impedance, E_{af}.

4.10 A 100 kVA, 208 V, 400 Hz, 8000 rev/min, three-phase, wye-connected synchronous generator supplies energy to a computer installation. The synchronous reactance, X_s, (also called series reactance) of the generator is 0.45 Ω. The stator resistance is 0.005 Ω. The three-phase computer load is 75 kW, 208 V, and 0.85 lagging power factor. (a) What is the generator current for rated computer load? (b) What is the line-to-line generator terminal voltage if the load circuit breaker is opened?

4.11 Make a plot of V_t vs. I_a for the generator of Problem 4.10 when operated as an isolated generator with a fixed dc field excitation and fixed shaft speed. Consider the load to be purely resistive and the dc field is set for rated stator current at rated terminal voltage. (Compare the result with Figure 4.23.)

4.12 Repeat Problem 4.11 for a 0.8 lagging power factor load.

4.13 Repeat Problem 4.11 for a 0.8 leading power factor load.

4.14 Repeat Example 4.4 for a decrease of 10 percent in parts *a* and *b*.

4.15 Sketch the stator flux density distribution corresponding to the three times given in Figure 4.19 except for an *acb* sequence. Compare the results of the *abc* and *acb* sequence and explain how to reverse the direction of rotation of a synchronous motor or an induction motor.

4.16 How do you change the speed of a synchronous motor?

4.17 How do you reverse the direction of rotation of a synchronous motor?

4.18 Explain what is meant by the expressions (a) pull-in torque? (b) pull-out torque?

4.19 What is the torque-speed curve for a synchronous motor, per se, supplied by a constant frequency power source and connected to a variable mechanical load?

4.20 Repeat Problem 4.4 for the machine connected as a motor.

4.21 Two synchronous generators are to be connected in parallel. The line voltage of each machine is set at 208 V. Determine the voltages across the synchronizing lamps connected in the "dark lamp" circuit as shown in Figure 4.24 for the following conditions (assume voltage $V_{ab} = 208\underline{/30°}$): (a) when the phase sequence of the two machines is the same and the voltage of machine A leads machine B by 30 degrees. (b) Same as part (a) except the voltages are in phase. (c) Same as part (a) except the voltages are 180° out of phase. (d) When the phase sequence of the two machines are opposite and the voltage V_{an} of Machine A is in phase with voltage $V_{a'n'}$ of Machine B. (e) Same as part (d) except the voltage of Machine A, (V_{an}), leads the voltage of Machine B, $(V_{a'n'})$, by 30°.

4.22 Describe two methods for starting a synchronous motor.

4.23 Describe the steps for synchronizing a 1000 MVA nuclear steam-turbine synchronous generator to a power system.

4.24 A synchronous motor is operating at leading power factor. This means that the field current is adjusted so that the stator current leads the terminal voltage. Neglect stator resistance and leakage reactance.

(a) Increase the field current by 10 percent without changing the load torque. Qualitatively, what changes occur in power output, in magnitude and phase of the stator current, and in magnitude of the torque angle? Explain by means of phasor diagrams representing the flux density and mmf waves.

(b) Increase the torque of the load by 10 percent without changing the field current. Explain the changes.

4.25 A synchronous motor is operating at rated load. An increase in its field excitation causes a decrease in stator current. Does the stator current lead or lag the terminal voltage?

4.26 A 500 hp, 60 Hz, 460 V, three-phase synchronous motor is operating at rated load and 0.7 leading power factor with a dc field current of 10 A. What will happen (qualitatively) to the line current if the dc field current is reduced to 3 A?

4.27 A 100 hp, 460 V, 60 Hz, three-phase, synchronous motor is operating from constant voltage mains at full load with a power factor of 0.80 leading. The operator slowly decreases the motor field current with the intention of ultimately making it zero. No other changes are made. Describe briefly the significant happenings as field current is decreased.

4.28 Electrical power is to be supplied to a three-phase 400 Hz system from a three-phase 60 Hz system through a motor-generator set consisting of two directly coupled synchronous machines. (a) What is the minimum number of poles which the motor may have? (b) What is the minimum number of poles which the generator may have? (c) At what speed in rev/min will the set specified in *a* and *b* operate?

4.29 What are the major advantages and disadvantages of a synchronous motor?

5

SINGLE-PHASE TRANSFORMERS

5.1 INTRODUCTION

The transformer is probably the most important device used for the efficient generation, transmission, and utilization of electrical energy. Transformers are used to step-up voltages, to step-down voltages, to give electrical isolation between circuits, to match impedances in electronic circuits, and to mix signals, to list a few purposes. See Figures 5.1-5.3 and 5.5-5.9 for examples. In large sizes the transformer has efficiencies in excess of 99.9 percent. The transformer has no moving parts to wear out, thus transformers built in 1903 are still in useful service.

Figure 5.1 Small dry-type single-phase transformer, 0.150 kVA, 50/60 Hz, 480/240:240/120 V. *(Courtesy of the Square D Company)*

Figure 5.2 Larger dry-type three-phase transformer, 45 kVA, 60 Hz, 480/240:240/120 V 3.4% Z. *(Courtesy of the Square D Company)*

A basic transformer consists of two windings electrically insulated from each other and wound on a common magnetic core. Generally, these two windings are wound one over the other. One winding is connected to an electrical source and is called the *primary winding*. The other winding is connected to a load and is called the *secondary winding*. The two windings usually have a different number of turns. The winding with the greatest number of turns is called the *high-voltage winding*. When the high-voltage winding is also the primary winding the transformer is connected as a *step-down transformer* since the output is at a lower voltage than the input. When the *low-voltage winding* is connected to the voltage source, the transformer is said to be a *step-up transformer*. The ratio of the number of primary winding turns to the secondary winding turns is called the transformation ratio and is designated by the letter *a*.

$$\frac{N_1}{N_2} = a \tag{5.1}$$

Figure 5.3 Dry-type transformer of Figure 5.2 with the cover removed. *(Courtesy of the Square D Company)*

If the coils are wound on a non-magnetic core, the assembly is referred to as an air core transformer. The air core transformer is extremely valuable in high frequency communication equipment. If the coils are wound on a ferromagnetic core, magnetic flux levels are attainable that allow for transfer of large quantities of electrical energy. For example, a transformer is used to step up the voltage of a 1200 MVA nuclear generator from 25 kV at 27,700 A to 765 kV at 906 A.

The emphasis in this text will be on power type transformers used mainly to change voltage levels in power distribution systems. It is important to recognize that there are transformers large enough to supply a city the size of Buffalo, New York. Also there are transformers so small that a dozen of them could fit in the palm of the hand.

5.2 IDEAL TRANSFORMER

The significant aspects of a transformer can be demonstrated by considering an ideal transformer. Figure 5.4 is a schematic diagram of a two-winding transformer wound on a ferromagnetic core. The properties of an ideal transformer are (1) the magnetic flux is confined to the core and windings, (2) the winding resistances are negligible, (3) the core losses are negligible, and (4) the permeability of the core is so high that only a negligible magnetomotive force is required to establish the core flux. These properties of ideal transformers are closely approached but never actually attained in real transformers.

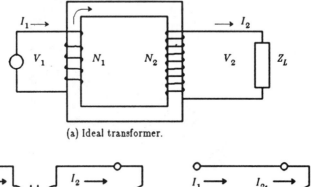

(a) Ideal transformer.

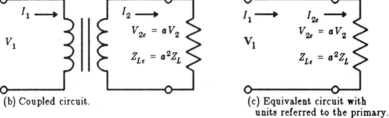

(b) Coupled circuit. (c) Equivalent circuit with units referred to the primary.

Figure 5.4 Ideal transformer circuit.

When an alternating voltage is supplied to the primary coil a magnetic field is established by Lenz's law, where

$$v_1 = N_1 \frac{d\phi}{dt} \qquad (5.2)$$

The changing flux causes a voltage in the secondary coil

$$v_2 = N_2 \frac{d\phi}{dt} \qquad (5.3)$$

Since the flux described in (5.2) and (5.3) is the same flux then

$$\frac{d\phi}{dt} = \frac{v_1}{N_1} = \frac{v_2}{N_2}$$

This leads to the expression that the coil voltages of a two winding transformer are directly related by the ratio of turns of each coil, where this ratio is noted by the letter a.

Figure 5.5 Exterior, open compartment view of a pad-mounted oil-type transformer rated 167 kVA, 12,470Y:240/120 V, 60 Hz. *(Courtesy of General Electric Company)*

$$\frac{v_1}{v_2} = \frac{N_1}{N_2} = a \tag{5.4}$$

When the ratio is greater than 1, the transformer is used to step down the voltage. When the ratio is less than 1, the transformer is used to step up the voltage. The primary coil magnetomotive force is defined as

$$F_1 = N_1 I_1 \tag{5.5}$$

and the secondary magnetomotive force is

$$F_2 = N_2 I_2 \tag{5.6}$$

The magnetomotive forces of (5.5) and (5.6) provide

$$F_1 = F_2 = N_1 I_1 = N_2 I_2$$

so that

$$\frac{I_1}{I_2} = \frac{N_2}{N_1} = \frac{1}{a} \tag{5.7}$$

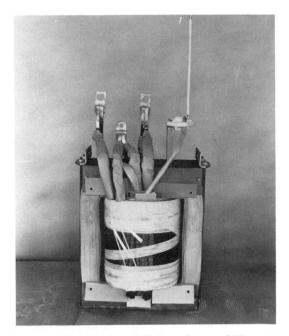

Figure 5.6 Interior view of the transformer of Figure 5.5. *(Courtesy of General Electric Company)*

In an ideal transformer all losses have been neglected so that

$$V_2 I_2 = \frac{V_1}{a}(a I_1) = V_1 I_1 \tag{5.8}$$

Thus, the input volt-amperes equals the output volt-amperes.

A simple circuit for an ideal transformer is shown in Figure 5.4b. In practice, it is desirable to develop an equivalent circuit or model as shown in Figure 5.4c, which does not include the coils. This can be accomplished by finding an output current that is called the equivalent secondary current, I_{2e}, and is equal to the input current. This would be:

$$I_{2e} = I_1 = \frac{I_2}{a}$$

Also, the equivalent output voltage would be:

$$V_{2e} = V_1 = a V_2$$

The effective impedance at the input of the transformer is different from the load impedance by the square of the turns ratio. Note that:

Figure 5.7 Exterior view of a three-phase, 150 kVA, 60 Hz, 12,000:480 V, load-center unit substation. *(Courtesy of General Electric Company)*

$$Z_{Le} = \frac{V_{2e}}{I_{2e}} = \frac{aV_2}{I_2/a} = a^2 \left(\frac{V_2}{I_2} \right) = a^2 Z_L \qquad (5.9)$$

Thus, a transformer may be used to match the impedance of devices, as described in Example 5.2.

In summary, for an ideal transformer the voltages are transformed in the direct ratio to the number of coil turns, the currents are transformed in inverse ratio to the number of coil turns, the impedances are transformed in the ratio squared of the number of coil turns, and the power and volt-amperes are unchanged between windings.

Example 5.1

A transformer with a 10:1 turns ratio and rated 50 kVA, 2,400:240 V, 60 Hz is used to step down the voltage of a distribution system. The secondary voltage is to be kept constant at 240 V.

a. What is the rated secondary current?

Figure 5.8 Assembled core and coils for an oil immersed, three-phase, power transformer rated 1000 MVA, 19,800:4330/2500 V, 60 Hz. *(Courtesy of General Electric Company)*

b. What load impedance will cause the rated current to flow?

c. What is the rated primary current?

d. What is the value of the load impedance referred to the primary side?

Solution

a. The rated secondary current is $50,000/240 = 208$ A.

b. The load impedance is $240/208 = 1.15$ Ω

c. Rated primary current is $50,000/2400 = 20.8$ A.

d. The load impedance referred to the primary coil is $(2400/240)^2(1.15) = 115$ Ω.

Example 5.2

An audio transformer is used to match the impedance of one channel of a 1000 Ω stereo amplifier to an 8 Ω speaker. The maximum transfer of power from the amplifier to the speaker occurs when the selected load

Figure 5.9 Winding a helical coil for a power transformer.
(Courtesy of General Electric Company)

impedance equals the amplifier (or source) impedance. Find the turns ratio of the audio transformer. See Figure 5.10.

Solution

By the maximum power transfer theorem,

$$Z_{(source)} = Z_{(load\ equivalent)}^*$$

or

$$1000 = a^2 8$$

Thus, the turns ratio a is

$$a = \sqrt{1000/8} = 11.2$$

5.3 PRACTICAL TWO-WINDING TRANSFORMER EQUIVALENT CIRCUITS

A practical transformer has losses. Thus, an equivalent circuit, Figure 5.11, which includes these losses, is obtained by taking the ideal transformer circuit and adding series and shunt impedances. The series resistances represent the winding losses (also called I^2R losses, or copper losses). The series reactances represent the leakage flux. This is the flux produced by

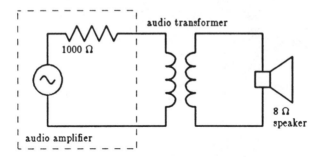

Figure 5.10 Impedance matching transformer for Example 5.2.

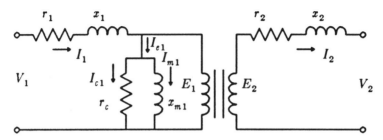

Figure 5.11 Transformer equivalent coupled circuit.

the current in the respective coil windings which is not common to the two windings. Hence, the names primary and secondary leakage flux, $\phi_{l\,1}$ and $\phi_{l\,2}$, as shown in Figure 5.12. The related reactances are $X_{l\,1}$ and $X_{l\,2}$. The magnetizing flux is represented by a shunt reactance, X_m, while the hysteresis and eddy current losses (these two losses are also called core losses or iron losses) are represented by the shunt resistance, r_c.

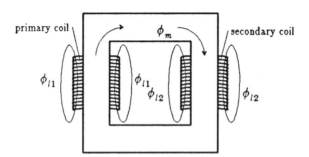

Figure 5.12 Transformer showing mutual and leakage fields.

The equivalent circuit of Figure 5.11 can be simplified by mathematically removing the coil and introducing equivalent values to produce the power-tee equivalent circuit of Figure 5.13a. A further simplification is obtained by moving the shunt leg ahead of the primary series elements as shown in Figure 5.13b to form the cantilever equivalent circuit. This is valid because the magnetizing current is small compared to the load current and is near a $90°$ phase shift from the load current. Note that these circuits apply to the step up as well as the step down connections.

Figure 5.13 has all quantities referred to the primary coil. It is possible to refer all quantities to the secondary coil as shown in Figure 5.14. The procedure for completing an equivalent circuit is given in Example 5.3.

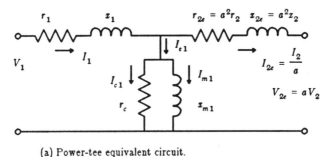

(a) Power-tee equivalent circuit.

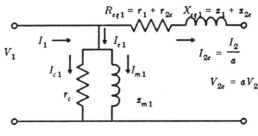

(b) Cantilever equivalent circuit.

Figure 5.13 Transformer equivalent circuits with units referred to the primary side.

5.4 VOLTAGE REGULATION AND EFFICIENCY

Most problems relating to transformers can be calculated to adequate accuracy by using the ideal transformer equivalent circuit. However, two computations are used to compare transformers. These are (1) voltage regulation and (2) efficiency, and can be obtained from the cantilever circuit.

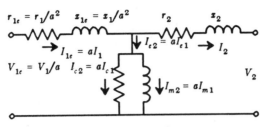

(a) Power-tee equivalent circuit.

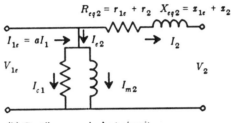

(b) Cantilever equivalent circuit.

Figure 5.14 Transformer equivalent circuits with units referred to the secondary side.

Voltage regulation is the change in the magnitude of load voltage between no load and full load. In equation form the voltage regulation, VR, is

$$VR = \frac{V_{2NL} - V_{2FL}}{V_{2FL}} \tag{5.10}$$

The no-load equivalent secondary voltage, V_{2eNL}, is the primary voltage V_1. Thus, for a transformer, (5.10) becomes

$$VR = \frac{V_1 - V_{2e}}{V_{2e}} \tag{5.11}$$

This means there is a need to determine the magnitude of the source voltage for rated secondary voltage. From the cantilever equivalent circuit, Figure 5.13b, the primary voltage is

$$\mathbf{V}_1 = \mathbf{V}_{2e} + \mathbf{I}_{2e}\mathbf{Z}_{eq} \tag{5.12}$$

Observe the phasor diagram for various load power factor angles in Figure 5.15. The relative size of the *IZ* phasor has been greatly exaggerated in order to show that the voltage regulation varies with the load power factor and that it becomes negative for large leading power factor angles.

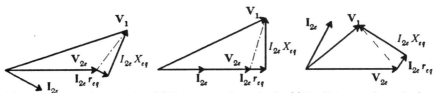

(a) Lagging power factor load.   (b) Unity power factor load.   (c) Leading power factor load.

Figure 5.15 Transformer phasor diagrams with IZ exaggerated for emphasis.

A second computation involves the finding of efficiency. Where efficiencies are near 100 percent, as in the case for large transformers, it is more precise to calculate in terms of the losses. Thus,

$$Eff = \frac{P_{out}}{P_{in}} = \frac{P_{in} - \text{losses}}{P_{in}} = 1 - \frac{\text{losses}}{P_{in}}$$

$$= 1 - \frac{\text{losses}}{P_{out} + \text{losses}} \tag{5.13}$$

Transformer losses include (1) core losses (hysteresis and eddy current losses) and (2) winding losses (I^2R or copper losses). For the cantilever equivalent circuit, the shunt resistance is directly related to the core losses. The core loss is obtained by measuring the power input when the transformer secondary is open-circuited.

Example 5.3

Given a 250 kVA, 4160:480 V, single-phase, 60 Hz transformer, the following parameters were obtained by test:

$$r_1 = 0.09 \ \Omega, \ \ X_1 = 1.7 \ \Omega, \ r_2 = 1.20 \times 10^{-3} \ \Omega,$$

$$X_2 = 2.26 \times 10^{-2} \ \Omega, \ r_{c1} = 31,600 \ \Omega, \ X_{m1} = j3240 \ \Omega.$$

Determine the following for the transformer connected step down. See Figure 5.16a.

(a) Calculate the primary voltage for rated load at 76 percent lagging power factor. The current I_2 for 76 percent power factor lags the secondary voltage by an angle

$$\theta = \arccos{(0.76)} = 40.54^{\circ}$$

$$I_{2 \text{ rated}} = \frac{250 \text{ kVA}}{480 \text{ V}} = 521 \text{ A}$$

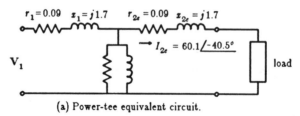

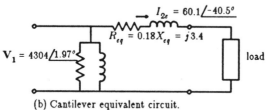

(a) Power-tee equivalent circuit.

(b) Cantilever equivalent circuit.

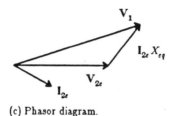

(c) Phasor diagram.

Figure 5.16 Circuits for Example 5.3a.

$$I_{2e \text{ rated}} = \frac{521}{(4160/480)} = 60.10 \text{ A}$$

$$\mathbf{V}_1 = \mathbf{V}_{2e} + \mathbf{I}_{2e}\mathbf{Z}_{eq}$$

$$= 4160\underline{/0^o} + (60.10\underline{/-40.54^o})(0.18 + j3.40)$$

$$= 4160\underline{/0^o} + (60.10\underline{/-40.54^o})(3.405\underline{/86.97^o})$$

$$= 4160\underline{/0^o} + 204.6\underline{/46.43^o}$$

$$= (4160 + j0) + (141.02 + j148.3)$$

$$= 4301 + j148.3 = 4303.6\underline{/1.97^o}$$

$$V_1 = 4304 \text{ V}$$

The phasor diagram is shown as Figure 5.15a.

    (b) Repeat (a) for 76 percent leading power factor ($\theta = 40.54^o$).

$$\mathbf{V}_1 = \mathbf{V}_{2e} + \mathbf{I}_{2e}\mathbf{Z}_{eq}$$

$$= 4160\underline{/0^\circ} + (60.10\underline{/+40.54^\circ})(0.18 + j3.40)$$

$$= 4160\underline{/0^\circ} + (60.1\underline{/+40.54^\circ})(3.405\underline{/86.96^\circ})$$

$$= (4160 + j0) + (-124.6 + j162.3)$$

$$= (4160\underline{/0^\circ}) + 204.6\underline{/127.5^\circ}) = 4035.4 + j162.3$$

$$V_1 = 4038 \text{ V}$$

(c) Calculate the transformer efficiency for part (a) and (b) with a core loss obtained from the no-load test at 547 W.

1. Efficiency for rated load at 76 percent lagging power factor.

Winding loss $= I_{2e}^2 R_{eq} = (60.0)^2(0.18) = 650.1$ W

Core loss = 547 W (given)

Total losses = 650 + 547 = 1197 W

$$P_{out} = VI \cos \theta = S \cos \theta$$

$$= (250 \text{ kVA})(0.76) = 190 \text{ kW}$$

$$Eff = 1 - \frac{\text{losses}}{P_{out} + \text{losses}} = 1 - \frac{1.197}{250(0.76) + 1.197}$$

$$= 1 - 0.00626 = 0.9937 = 99.37\%$$

2. The efficiency for 76 percent leading power factor is the same as for part (c)1.

(d) Calculate voltage regulation for parts (a) and (b).

(1) $VR = \dfrac{4304 - 4160}{4160} = 0.0346$ per unit = 3.46 percent

(2) $VR = \dfrac{4038 - 4160}{4160} = -0.0293$ per unit = -2.93 percent

---

5.5 CONSTRUCTION

A. Arrangement of Core and Windings

A transformer constructed with the core outside the windings is called a shell type as shown in Figure 5.17a. A transformer constructed so that the winding surrounds the core is called a core type, as shown in Figure 5.17b.

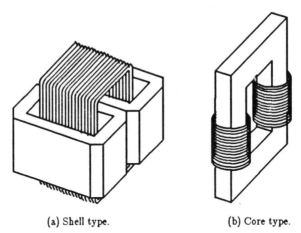

(a) Shell type.   (b) Core type.

Figure 5.17 Single-phase transformer construction.

B. Type of Core

Ferromagnetic cores may be made from (a) laminated sheets (also called stacked punchings), (b) wound cores, and (c) compressed powder cores. The magnetic material used in making the cores is discussed in Chapter 2.

C. Methods of Cooling

Almost all small communication and control transformers and some indoor-type power transformers are air cooled. Air-cooled transformers used for power distribution in buildings are available in sizes from 0.05 to 750 kVA, as shown in Figure 5.1-5.3. These are referred to as dry-type transformers. Most outdoor transformers are filled with an insulating oil that greatly increases the cooling efficiency. When these oil-type transformers are used in buildings, they must be housed in special fire-retaining rooms or vaults. The fire risk was greatly reduced by using an askarel type oil, also called PCB. This type of oil is nonbiodegradable and has been reported to be a cancer-causing substance. In recent years, a silicon-type oil has been introduced on the market that may well serve the purpose for nonflammable oil.

The distribution and transmission transformers may be further categorized by the method of supplementary cooling. Some transformers have fins that increase the effective cooling, as shown in Figure 5.7. Electric fans may be added that increase the cooling. When circulating oil pumps are added, the transformer is said to be forced-oil cooled. For some types of transformers, water pipes are extended through the oil and are called water-cooled transformers. In the larger transformers, the transformer may be operated with several types of cooling. This has lead to what is called triple-rated transformers. For example, a transformer may be self-cooled operated to 12 MVA with Oil Air (OA) type of cooling. The transformer rating may be extended to 16 MVA by turning on fans for Forced Air Cooling (FA), and then further extended to 20 MVA by turning on the oil pumps for Forced Oil Air (FOA) cooling. This transformer is said to be rated 12/16/20 MVA and is designated OA/FA/FOA. Another example of a cooling system has two stages of Forced Water (FW) cooling and is designated OA/FW/FW.

D. Polarity Marking and Multiple Windings

ANSI Markings. The industry and users of power transformers have arrived at a standard of designating transformer terminals. The high-voltage terminals are designated H1, H2, etc., and the low voltage terminals as X1, X2, etc., as shown in Figure 5.18. The H1 and X1 indicate polarity markings so that when the current enters at terminal H1, the current leaves X1. These have the same meaning as the *dot marks* used in many circuit texts in describing magnetically coupled circuits. The terminals may designate several taps or several coils as shown in Figure 5.18.

The rating of a transformer includes voltage, frequency, and kVA. The kVA rating is the kVA output that the transformer can deliver at rated voltage and frequency under usual service conditions without exceeding the standard limits of temperature increase. Thus, a 60 Hz, 100 kVA, 2400:240 V transformer, when operated at rated full load as a step-down transformer, would be delivering 100 kVA at 240 V for a frequency of 60 Hz. The source voltage will be slightly different from 2400 V depending on the magnitude and power factor of the load. If this transformer were operated as a step-up transformer, it would be operating at its rated value when it is delivering 100 kVA, 2400 V and 60 Hz.

The transformers discussed so far have been of the two-winding type. Some distribution transformers are made with four windings on a common core as shown in Figure 5.19. The windings are constructed so that a choice may be made between connecting the coils in series or parallel. The standard notation for designating a transformer includes kVA and voltage ratings. A slant signifies multiple voltage designation, and a colon separates the high-voltage and low-voltage coils. For example, the transformer of Figure 5.19 is listed as 25 kVA and 480/240:240/120 V.

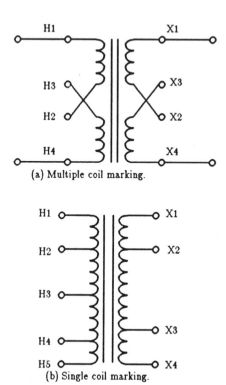

(a) Multiple coil marking.

(b) Single coil marking.

Figure 5.18 Transformer terminal markings.

5.6 TRANSFORMER TESTS (Advanced Topic)

The numerical values for the transformer equivalent circuit parameters may be determined from performance tests. The two important tests are called the (1) *no-load test* and (2) the *load-loss test*. For many years these two tests were referred to as the *open-circuit* and *short-circuit* tests, respectively. However, the ANSI standard for the open-circuit test is to apply rated voltage to one transformer coil while measuring the voltage of the second coil; it is used to determine the turns ratio of the transformer. Likewise a short-circuit test should properly refer to the test performed to determine the mechanical strength of the transformer and is conducted by applying rated voltage for a very short duration to one coil of the transformer while the other coil is shorted. These name changes are recent IEEE and ANSI changes so that most literature still uses the titles of open-circuit and short-circuit tests for the no-load test and load-loss tests, respectively.

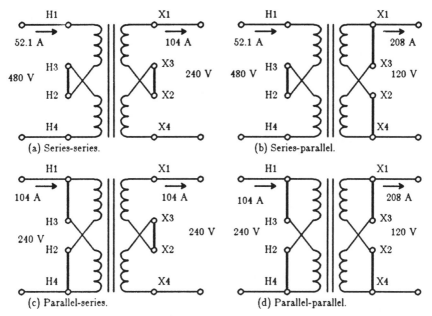

Figure 5.19 Possible connections for a four-coil transformer rated 25 kVA, 480/240:240/120 V.

A. No-load Test

The no-load test is performed by applying rated voltage to one coil of the transformer. The source is usually applied to the low-voltage coil since a lower voltage source is generally more readily accessible. The input voltage, current, and power is measured. The power is a measure of the core loss (hysteresis and eddy current losses) plus a small amount of the winding I^2R loss in the connected coil. For power size transformers the winding loss can generally be neglected.

The shunt admittance is determined from

$$Y_o = \frac{I_{NL}}{V_{NL}} \tag{5.14}$$

and

$$\cos \theta_o = \frac{P_{NL}}{V_{NL} I_{NL}} \tag{5.15}$$

The shunt resistive component is determined from

$$G_c = \frac{P_{NL}}{V_{NL}^2} \qquad (5.16)$$

and the magnetizing susceptance is

$$B_m = \sqrt{Y_o^2 - G_c^2} \qquad (5.17)$$

The values for the shunt leg impedance are normally given as

$$R_c = \frac{1}{G_c} \qquad (5.18)$$

and

$$X_m = \frac{1}{B_m} \qquad (5.19)$$

Note that the values obtained from (5.18) and (5.19) are not the same as for a series-type of magnetization circuit where the impedance is:

$$Z_{(o\ ser)} = R_{(c\ ser)} + jX_{(m\ ser)} = \frac{1}{Y_o} = \frac{1}{G_c - jB_m} \qquad (5.20)$$

where

$$R_{(c\ ser)} = \frac{G_c}{G_c^2 + B_m^2} \qquad (5.21)$$

and

$$X_{(m\ ser)} = \frac{B_m}{G_c^2 + B_m^2} \qquad (5.22)$$

The values for R_c and X_m are different when referenced to the primary or the secondary coils. It may be wise to add notation indicating whether Y_o refers to the primary coil, Y_{o1}, or the secondary coil, Y_{o2}. The significance of this concept should become obvious from Example 5.4.

B. Load-loss Test

The load-loss test is performed by applying a reduced voltage to one coil of the transformer while the other coil is shorted. Recall that the high-voltage coil requires less current; therefore, the test source has less current demand if the high-voltage coil is connected to the power source for the load-loss test. The transformer equivalent series impedance for the cantilever equivalent circuit is therefore:

$$Z_{eq} = \frac{V_{LL}}{I_{LL}} \tag{5.23}$$

and

$$\cos\theta_{eq} = \frac{P_{LL}}{V_{LL}I_{LL}} \tag{5.24}$$

Thus

$$\mathbf{Z}_{eq} = Z_{eq} \underline{/\theta} = R_{eq} + jX_{eq} \tag{5.25}$$

Another method for obtaining the values for the equivalent series resistance is

$$R_{eq} = \frac{P_{LL}}{I_{LL}^2} \tag{5.26}$$

and the equivalent series reactance is

$$X_{eq} = \sqrt{Z_{eq}^2 - R_{eq}^2} \tag{5.27}$$

Example 5.4

Determine the cantilever equivalent circuit of a transformer rated at 1000 kVA, 66,000:6600 V, 60 Hz with units referred to the HV coil as primary. The no-load test measured on the low-voltage coil is: 6600 V, 4.0 A, and 9000 W. The load-loss test measured on the high voltage coil is: 3500 V, 16 A, and 8000 W.

Solution

The series impedance is found from the load-loss test data

$$Z_{eq1} = \frac{V_{LL}}{I_{LL}} = \frac{3500}{16.0} = 218.8 \ \Omega$$

$$\cos\theta = \frac{P_{LL}}{V_{LL}I_{LL}} = \frac{8000}{3500\times16} = 0.1429$$

$$\theta = 81.79°$$

Thus

$$Z_{eq1} = 218.8\underline{/81.79°} = 31.25 + j216.5$$

The shunt admittance is found from the no-load test data.

$$Y_2 = \frac{I_{NL}}{V_{NL}} = \frac{4.0}{6600} = 0.606\times10^{-3}\text{ S}$$

$$\cos\theta = \frac{P_{NL}}{V_{NL}I_{NL}} = \frac{9000}{6600\times4} = 0.3409$$

$$\theta = 70.07°$$

$$Y_2 = 0.606\times10^{-3}\underline{/-70.07°} = (0.2066 - j0.5693)\times10^{-3}\text{ S}$$

Note that Y_2 is in units referred to the low-voltage coil. Divide Y_2 by a^2 to obtain Y_1, which is the shunt admittance referred to the high voltage coil. Thus

$$Y_1 = \frac{Y_2}{a^2} = (6.061\times10^{-6})\underline{/-70.07°}\text{ S}$$

The values of the shunt circuit in ohms is

$$R_c = 484,000\ \Omega \text{ and } X_m = j175,000\ \Omega$$

The cantilever equivalent circuit with the parameters appropriately labeled is shown in Figure 5.20.

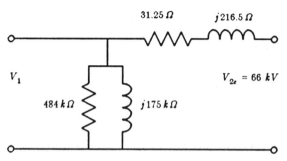

Figure 5.20 Cantilever equivalent circuit for Example 5.4.

Example 5.5

Determine the power-tee equivalent circuit for the transformer of Example 5.4. The resistances of a well-designed transformer are such that

$$r_1 = r_{2e} = a^2 r_2$$

Thus,

$$r_1 = r_{2e} = \frac{31.25}{2} = 15.63$$

The reactances are such that

$$X_1 = X_{2e} = a^2 X_2$$

Thus,

$$X_1 = X_{2e} = \frac{216.5}{2} = 108.3$$

The power-tee equivalent circuit with the parameters appropriately labeled is shown as Figure 5.21. The secondary impedances are

$$r_2 = \frac{r_{2e}}{a^2} = \frac{15.6}{100} = 0.156$$

$$X_2 = \frac{X_{2e}}{a^2} = \frac{108}{100} = 1.08$$

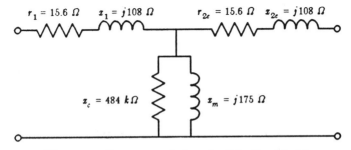

Figure 5.21 Power-tee equivalent circuit for Example 5.5.

5.7 PER UNIT QUANTITIES

Users and suppliers of transformers refer to the *percent impedance of a transformer*. This expression refers to the series impedance of the cantilever equivalent circuit measured as a percentage of the transformer rating and allows for a convenient designation and procedure for calculating voltage regulation and short circuit currents. When referring to a given transformer or machine, the base is normally considered the rated kVA and rated voltages of the device. There will be a different reference of base value for the high-voltage and low voltage windings. The reference value based on device ratings are therefore:

S_{base} = defined volt-ampere reference which is the same for the HV and LV coils

V_{base} = defined voltage reference

I_{base} = defined current reference = $\dfrac{S_{base}}{V_{base}}$

$$Z_{base} = \frac{V_{base}}{I_{base}} = \frac{V_{base}}{S_{base}/V_{base}}$$

$$= \frac{(V_{base})^2}{S_{base}} = \frac{(kV_{base})^2}{MVA_{base}} = \frac{(kV_{base})^2 \times 1000}{kVA_{base}} \tag{5.28}$$

The relationships between actual (or measured quantities) are

$$V_{\text{per unit}} = \frac{V_{meas}}{V_{base}}$$

$$I_{\text{per unit}} = \frac{I_{meas}}{I_{base}}$$

$$Z_{\text{per unit}} = \frac{Z_{meas}}{Z_{base}} \tag{5.29}$$

The following example should demonstrate the procedure.

Example 5.6

Given the transformer of Example 5.3, calculate (a) the percent impedance and (b) the percent exciting current.

Solution

The measured series equivalent impedance referred to the high-voltage coil of the transformer is

$$r_{eq1} = r_1 + r_{2e} = r_1 + a^2 r_2$$
$$= 0.09 + (8.667)^2 1.20 \times 10^{-3} = 0.09 + 0.09 = 0.18$$

$$x_{eq1} = x_1 + x_{2e} = x_1 + a^2 x_2$$
$$= 1.70 + (8.667)^2 2.26 \times 10^{-2} = 1.70 + 1.70 = 3.40$$

$$Z_{eq1} = r_{eq1} + x_{eq1}$$
$$= 0.18 + j3.40 = 3.4 \underline{/86.97^\circ}$$

The base values for the high-voltage coil of the transformer are

$$S_{base\ HV} = S_{base\ LV} = 250 \text{ kVA} = 0.250 \text{ MVA}$$

$$V_{base\ HV} = 4160 \text{ V}$$

$$Z_{base\ HV} = \frac{(kV_{base\ HV})^2}{MVA_{base}} = \frac{(4.160)^2}{0.250} = 69.22 \ \Omega$$

The base values for the low-voltage coil of the transformer are

$$V_{base\ LV} = 480 \text{ V}$$

$$Z_{base\ LV} = \frac{(0.480)^2}{0.250} = 0.9216 \ \Omega$$

The per unit values referred to the HV coil are

$$Z_{eq1\ per\ unit} = \frac{Z_{eq1}}{Z_{base\ HV}}$$

$$= \frac{0.18 + j3.4}{6.922} = (0.0026 + j0.0491) \text{ per unit } \Omega$$

The per unit values referred to the LV coil are

$$r_{eq2} = \frac{r_1}{a^2} + r_2 = \frac{0.9}{(8.667)^2} + 0.00120 = 0.00240$$

$$x_{eq2} = \frac{x_1}{a^2} + x_2 = \frac{1.70}{(8.667)^2} + 0.0226 = 0.0452$$

$$Z_{eq2} = 0.0024 + j0.0452 = 0.0453 \underline{/86.96^\circ}$$

$$Z_{eq2 \text{ per unit}} = \frac{Z_{eq2}}{Z_{base \, LV}}$$

$$= \frac{0.0024 + j0.0452}{0.9216} = 0.00260 + j0.0491 \text{ per unit } \Omega$$

Note that the per unit impedance is the same for either coil. This is so because the MVA base is the same for either side of the transformer and the KV base values are related by the turns ratio of the transformer. The more common method for using per unit values is that given in Example 5.7.

Example 5.7

A single-phase transformer nameplate includes the following data: 75 kVA, 2400:120 V, 2.1% impedance. What is the magnitude of the secondary short circuit current when rated voltage is applied to the primary coil?

Solution

Method 1. The 2.1% impedance gives reference to the cantilever series equivalent impedance with the transformer rating as the base. Thus,

$$S_{base} = 75 \text{ kVA}$$

$$V_{baseHV} = 2400 \text{ V}$$

and

$$Z_{base1} = \frac{(kV_{base})^2 \times 1000}{kVA_{base}} = \frac{(2.4)^2 1000}{75} = 76.8 \text{ per unit } \Omega$$

$$Z_{eq1} = Z_{\text{per unit}} Z_{base1} = 0.021 \times 76.8 = 1.613 \ \Omega$$

The short circuit current using the cantilever equivalent circuit of Figure 5.13b is

$$I_{2e} = \frac{V_1}{Z_{eq1}} = \frac{2400}{1.613} = 1488 \text{ A}$$

The secondary current is

$$I_2 = aI_{2e} = (2400/120)(1488) = 29{,}760 \text{ A}$$

Note that any equipment such as fuses, circuit breakers, panels, etc. located near the transformer secondary must be capable of carrying for a short time and interrupting almost 30,000 A. This is referred to as the AIC (amperes interrupting capacity) of the connected equipment.

Method 2. The per unit short circuit current is

$$I_{1 \text{ per unit}} = \frac{V_{1 \text{ per unit}}}{Z_{\text{per unit}}} = \frac{1.0}{0.021} = 47.62 \text{ per unit A}$$

$$I_{2 \text{ base}} = \frac{kVA_{base}}{kV_{2 \text{ base}}} = \frac{75}{0.120} = 625 \text{ A}$$

Note that $I_{2 \text{ per unit}} = I_{1 \text{ per unit}}$ then

$$I_2 = (I_{2 \text{ per unit}})(I_{2 \text{ base}}) = (47.62)(625) = 29{,}760 \text{ A}$$

This second method can be faster when the per unit ideas are fully understood.

On occasion it is desirable to use a base other that the device rating. The equations for converting the impedance from one base to another is

$$\frac{Z_{\text{per unit }new}}{Z_{\text{per unit }old}} = \frac{Z_{meas}/Z_{base\ new}}{Z_{meas}/Z_{base\ old}}$$

$$= \frac{Z_{base\ old}}{Z_{base\ new}} = \frac{(kV_{base\ old})^2/MVA_{base\ old}}{(kV_{base\ new})^2/MVA_{base\ new}}$$

$$\frac{Z_{\text{per unit }new}}{Z_{\text{per unit }old}} = \left(\frac{kV_{base\ old}}{kV_{base\ new}}\right)^2 \left(\frac{MVA_{base\ new}}{MVA_{base\ old}}\right) \tag{5.30}$$

Example 5.8

What is the percent impedance for the transformer of Example 5.7 when used in a distribution system where it is desirable to use an S_{base} of 2000 kVA and a V_{base} at the high-voltage coil of 2200 V.

Solution

Use (5.30) to determine the new per unit impedance. Observe that the high-voltage coil is rated 2400 V.

$$Z_{\text{per unit } new} = (0.021)\left(\frac{2400}{2200}\right)^2\left(\frac{2000}{75}\right)$$

$$= 0.666 \text{ per unit } \Omega = 66.6 \text{ percent } \Omega$$

It is possible to solve this problem by referring the quantities to the secondary coil. The new high-voltage coil base voltage is

$$V_{2\ base\ new} = \frac{1}{a}V_{1\ base\ new} = (120/2400)(2200) = 110 \text{ V}$$

Then the new percent impedance is

$$Z_{\text{per unit } new} = (0.021)\left(\frac{120}{110}\right)^2\left(\frac{2000}{75}\right) = 0.666 \text{ per unit } \Omega = 66.6 \text{ percent } \Omega$$

5.8 AUTOTRANSFORMERS

An autotransformer is a transformer that has part of the winding common to the primary and secondary circuits as shown in Figures 5.22 and 23. These are often more efficient and less expensive to build. This is especially true where the voltage ratios are close to unity. The autotransformer is used in power transmission systems, as voltage regulators, as fluorescent lamp ballasts, and for variable ac power sources (manufactured under trade names such as "Variac," "Powerstat," and "Voltstat.")

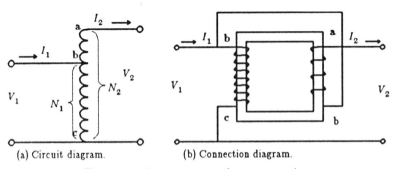

(a) Circuit diagram.   (b) Connection diagram.

Figure 5.22 Step-up autotransformer connection.

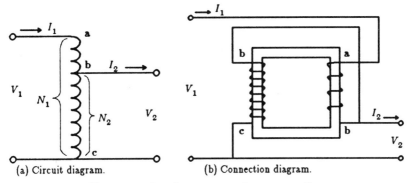

(a) Circuit diagram.   (b) Connection diagram.

Figure 5.23 Step-down autotransformer connection.

A. Equations for Ideal Autotransformers

The primary and secondary voltages are related by the number of turns common to the respective voltages as shown in Figure 5.22 and 5.23 so that

$$\frac{V_1}{V_2} = \frac{N_1}{N_2} = a \tag{5.31}$$

The line currents are related by the inverse of the turns ratio, or

$$\frac{I_1}{I_2} = \frac{N_2}{N_1} = \frac{1}{a} \tag{5.32}$$

The current in the common leg is simply the sum of the currents at the junction. Proving these equations is assigned as homework.

Example 5.9

Consider the 100:80 V step-down autotransformer of Figure 5.24a with a secondary load current of 15 A. The turns ratio is

$$a = \frac{N_1}{N_2} = \frac{V_1}{V_2} = \frac{100}{80} = 1.25$$

The primary line current for a secondary load current of 15 A is

$$I_1 = \frac{I_2}{a} = \frac{15}{1.25} = 12 \text{ A}$$

The current in the common coil is

$$I_c = I_1 - I_2 = 12 - 15 = -3 \text{ A}$$

with the current up in the common coil.

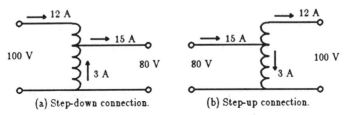

(a) Step-down connection.   (b) Step-up connection.

Figure 5.24 Autotransformer circuit for Examples 5.9 and 5.10.

Example 5.10

Repeat Example 5.9 for a step-up autotransformer circuit of Figure 5.24b with a 12 A load and a voltage ratio of 80 to 100 V. Thus, the turns ratio is a $N_1/N_2 = 80/100 = 0.8$. The primary line current for a secondary load current of 12 A is

$$I_1 = \frac{I_2}{a} = \frac{12}{0.8} = 15 \text{ A}$$

The current in the common coil is then

$$I_c = 15 - 12 = 3 \text{ A}$$

with the current down in the common coil.

Example 5.11

A 7200:277 V, 75 kVA distribution transformer is to be considered for use as a step-down autotransformer by connecting the 7200 V and 277 V coils in series (see Figure 5.25).

(a) What are the voltage, current, and kVA ratings when the transformer is connected as a two-winding transformer?

(b) What are the voltage, current, and kVA ratings when the transformer is connected as the autotransformer?

(c) Is there a problem if the given distribution transformer was specified as a 7200Y/12470:277Y/480 transformer?

Solution

a) The two-winding transformer ratings are

High Voltage Winding	Low Voltage Winding
7200 V	277 V
10.4 A	271 A
75 kVA	75 kVA

(b) The autotransformer ratings are

$$V_{in} = 7200 + 277 = 7477 \text{ V}$$

$$V_{out} = 7200 \text{ V}$$

The current rating of high-voltage coil is

$$I_{H\ rated} = \frac{75000}{7200} = 10.42 \text{ A}$$

The current rating of the low-voltage coil is

$$I_{X\ rated} = \frac{75000}{277} = 271 \text{ A}$$

The rated input current is limited to the rating of the 277 V coil, or 271 A. Therefore, the output current is

$$I_{out} = (7477/7200)271 = 281.4 \text{ A}$$

This means the current in the common leg is

$$I_c = I_{out} - I_{in} = 281.4 - 271 = 10.4 \text{ A}$$

which is the rating of the 7200 V coil. The kVA rating of the transformer connected as an autotransformer is

$$kVA_{in} = 7477(271) = 2026 \text{ kVA}$$

$$kVA_{out} = 7200(281.4) = 2026 \text{ kVA}$$

This means that a 75 kVA, 7200:277 V two-winding transformer could be used as an autotransformer to transform 2026 kVA at 7477:7200 V.

(c) There is a caution that must be observed when using any two-winding transformer as an autotransformer. Many transformers are designed for use as part of a three-phase wye bank. These transformers have limited insulation to ground. If used as autotransformers, the metal transformer case is at or near the high-voltage value and a safety hazard exists. Thus, a transformer labeled 7200Y/12470:277Y/480 is a 7200:277 V transformer designed for use as one transformer of a three-phase, wye-wye connected 12,470:480 V system. This transformer could not be used as part of an autotransformer system unless the transformers were mounted on an insulated structure and kept in a vault so that they could not be touched by personnel. By changing the method of insulation it would be possible to make an autotransformer rated 2026 kVA, 7477:7200 V.

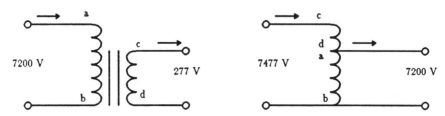

Figure 5.25 Circuit for Example 5.11

5.9 INSTRUMENT TRANSFORMERS

A. Potential Instrument Transformers

This type of transformer is designed for metering and relaying purposes. Usually, it is used for coupling a 120 V voltmeter to a high voltage system. These transformers are designed for a minimum magnetizing current and for low values of series impedance in order to give an accurate voltage ratio.

B. Current Transformers

The current transformer is used for metering or relaying purposes. Many current transformers (CT's) are shaped like a doughnut or toroid (Figure 11.13) where the primary winding is the conductor carrying the circuit current, which is led through the center of the toroid. For utility-type current transformers, the secondary is rated at 5 A for use with 5 A meters and relays. For example, a CT for measuring the current in a conductor carrying 1000 A would have a turns ratio of 200 with a secondary current of 5 A.

There is a major caution required when using CT's. The secondary must always be connected to a low impedance load or else shorted. Because of the removal of the demagnetizing effect of the secondary current, the flux then rises to a value determined by the primary current. Since the current transformer is normally operated at low flux density, the flux may increase considerably until it is limited by saturation, and the rms value of the voltage across the open secondary terminals may rise to several hundred volts. Because of the distortion due to the core saturation, the induced voltage has a peaked wave form, and the peak may be sufficiently high to be dangerous to life and to the transformer insulation.

PROBLEMS

5.1 In reference to a transformer, what is the meaning of the expression (a) primary winding? (b) secondary winding?

5.2 A transformer has 800 turns on the primary winding and 160 turns on the secondary winding. The secondary is rated at 37.5 kVA and 480 V. Determine (a) the transformer turns ratio, (b) the approximate primary voltage, (c) the rated full load secondary current, (d) the rated full load primary current with exciting current neglected, and (e) the primary kVA rating.

5.3 Given a 120:18 V *model car* power supply transformer with the secondary rated at 2.1 A, (a) calculate the transformer turns ratio; (b) approximate the primary current for rated secondary current; (c) determine the volt-ampere rating of the transformer.

5.4 Can a 240:120 V transformer be operated with the 120 V coil as primary?

5.5 What is meant by transformer *rated kVA ?*

5.6 Can transformers be operated at voltages other than nameplate voltages? Explain.

5.7 What is meant by hysteresis loss?

5.8 What are eddy currents, and how are they minimized?

5.9 (a) Why should dry type transformers never be overloaded? (b) Can an oil-filled transformer be overloaded?

5.10 Locate a transformer, record the information shown on the nameplate, and discuss the meaning of the information.

5.11 Given a 7620:240/120 V, 167 kVA, oil-filled distribution transformer, calculate the primary voltage when the transformer secondary is operated at rated current, at the 240 V connection, and at an 80 percent lagging power factor. The equivalent cantilever circuit of Figure 5.26 was calculated from manufacturer's published data.

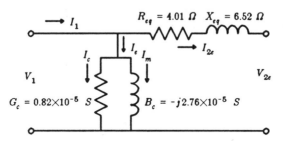

Figure 5.26 Equivalent cantilever circuit for Problem 5.11.

5.12 Repeat Problem 5.11 for a unity power factor load.

5.13 Repeat Problem 5.11 for a 60 percent leading power factor load.

5.14 Calculate the efficiency of the transformer of Problem 5.11 for a rated transformer load at 80 percent lagging power factor. The no load loss as determined from the no load test is 476 W. ($P = V^2 G = (7620)^2(0.82 \times 10^{-5}) = 476$ W.)

5.15 Repeat Problem 5.14 for unity power factor.

5.16 Repeat Problem 5.14 for 60 percent leading power factor.

5.17 Determine the cantilever equivalent circuit [(a) series impedance, and (b) shunt impedance] for a 75 kVA, 7200:240/120 V, 1045 lb, oil-filled distribution transformer. The following test data were obtained: (1) open circuit test (with voltage applied to the low-voltage coils connected in series): 240 V, 4.69 A, 240 W; (2) short circuit test (with voltage applied to the high voltage coil): 133 V, 10.1 A and 295 W.

5.18 Calculate the efficiency for the transformer of Problem 5.17 for a unity power factor load at rated secondary voltage and rated secondary current.

5.19 Calculate the efficiency for the transformer of Problem 5.17 for (a) a load of 0.8 power factor lagging and (b) a load of 0.6 power factor leading.

5.20 The equivalent series impedance referred to the high voltage side of a single-phase 46,000:4160 V, 5000 kVA transformer is $(0.73 + j2.10)$ Ω. Calculate the magnitude of the primary voltage when the transformer is operating at rated secondary voltage, at rated secondary current, and 0.53 leading power factor.

5.21 The equivalent series impedance referred to the high voltage side of a single-phase 138,000:69,000 V, 50 MVA transformer is $(4.28 + j10.59)$ Ω. Calculate the primary voltage for rated secondary current at rated secondary voltage and at 0.6 leading power factor.

5.22 Can transformers be used in parallel? Explain.

5.23 What is meant by voltage regulation of a transformer?

5.24 Two separate transformers are used to couple an 8 Ω loudspeaker to a stereo amplifier channel. The equivalent circuit for each of the amplifier channels is a 500 Ω resistance in series with a voltage source. What is the approximate turns ratio of each of the transformers for maximum power transfer?

5.25 A 150,000 kVA, 138:115 kV, single-phase autotransformer is used in a step-down connection. Determine the (a) primary current, (b) secondary current, and (c) current in the common coil.

5.26 It is given that a 12,470:480 V, 167 kVA single-phase transformer has a series impedance of $(0 + j18.6)$ Ω when referred to the primary coil with the transformer used in a step-down connection. (a) What is the rated primary current? (b) What is the rated secondary current? (c) If the transformer secondary terminals were accidentally shorted, what current would flow through the (1) secondary coil and (2) primary coil if it is assumed that the transformer is supplied from a constant 12,470 V source?

5.27 Repeat Problem 5.26 for a series impedance of $(0 + j9.31)$ Ω when referred to the primary coil.

5.28 The transformer of Problem 5.26 supplies a single-phase load with #12 AWG copper conductor 100 feet from the transformer. A 20 A circuit breaker protects the conductor against overload. What current must the circuit breaker be able to interrupt for a short circuit on the load side of the breaker if it is found that the impedance of the conductor is $(1.62 + j0.1)$ Ω per 1000 feet?

5.29 What is a potential transformer used for?

5.30 What is a current transformer? Discuss a major caution regarding the use of current transformers.

5.31 Given a 150 kVA, 7200:480 V, single-phase transformer with a series impedance of $(0.71 + j6.91)$ Ω with the impedance referred to the high-voltage coil. (a) Calculate the secondary current for a load of $(1.3 + j0.8)$ Ω at rated secondary voltage. (b) Calculate the magnitude of the primary voltage required to produce the load current of part (a). (c) What is the secondary current when the secondary terminals are shorted and the rated voltage is connected to the primary high-voltage terminals?

5.32 Given a 25 kVA, 480:240 V, single-phase transformer with an internal series impedance of $(0.02 + j0.18)$ Ω with the impedance referred to the high-voltage coil. If a short (or fault) occurs at point F of Figure 5.27, what magnitude of current must the 50 A circuit breaker interrupt? Neglect the impedance of the conductor between the transformer and fault.

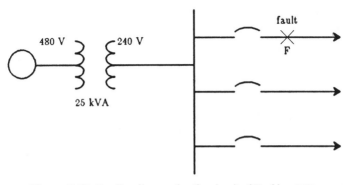

Figure 5.27   One-line diagram for the circuit of Problem 5.32.

5.33 Table 5.1 lists the no-load and load-loss test data for a number of transformers. Determine the equivalent cantilever circuit assuming the high-voltage coil is the primary coil and that the values are to be referred to the primary coil for each transformer.

	Rating			No-load Test Measured on LV Coil			Load-loss Test Measured on HV Coil		
	kVA	Voltages	f	V	I	P	V	I	P
A	10	252400:120	60	120	1.0	75	120	4.5	250
B	10	2400:240/120	60	240	0.46	63	43.6	4.3	125
C	20	2400:240	25	240	2.0	150	240	9.0	450
D	100	11,000:2200	50	2200	1.5	1000	600	10.0	1200
E	1000	66,000:6600	60	6600	4.0	9000	3500	16.0	8000
F	5000	15,000:4000	60	4000	60.0	30,000	1500	360	4000

Table 5.1   Two-winding Transformers

5.34 Table 5.1 lists the no-load and load-loss test data for a number of transformers. Determine the equivalent cantilever circuit assuming the low-voltage coil is the primary coil and that the values are to be referred to the primary coil for each transformer.

5.35 Calculate the voltage regulation and efficiency for the transformers in Table 5.1, assuming each is operated step-down at its rated load with a 0.85 lagging power factor and rated secondary voltage.

5.36 Determine the percent series impedance for each transformer in Table 5.1.

5.37 Find the equivalent input impedance for the circuit of Figure 5.28.

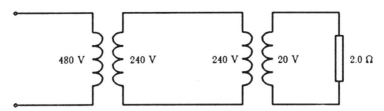

Figure 5.28 Equivalent circuit for Problem 5.37.

5.38 Verify equations (5.31) and (5.32).

6

THREE-PHASE
TRANSFORMER
CONNECTIONS

6.1 STANDARD THREE-PHASE CONNECTION

Electrical transmission and distribution systems are three-phase systems operating at several voltage levels. The generators are usually located at great distances from metropolitan areas and produce electrical energy at a potential of 13 kV to 25 kV. A "generator unit transformer" then steps the voltage to transmission levels of 220 kV to 765 kV. The electrical energy is transmitted to metropolitan areas often over distances of hundreds of miles. The voltage is then reduced or "stepped down" to values between 46 kV and 138 kV for sub-transmission about the community. Then for distribution within neighborhoods the voltage is further reduced to levels between 4.16 kV and 34.5 kV. Finally, the voltage is reduced to 120/240 single-phase, 120/240 three-phase, or 120/208 three-phase for residential or light commercial use. Commercial and industrial customers may receive energy at other appropriate values of three-phase voltages.

These various transformation levels require three-phase transformer connections, of which there are four standard types. These are

1. wye-wye
2. delta-wye
3. wye-delta
4. delta-delta

A. Calculations

The details of the voltage and current values are best understood by solving problems involving the various connections. The calculations generally neglect losses and exciting current so that the transformers are considered as ideal transformers on the basis of voltage and current ratios only. There are two common types of diagrams: the *connection diagram,* which is useful when wiring transformers; and the *circuit diagram,* which is useful when solving computational problems. Two of the four standard connections are shown in Figures 6.1 and 6.3. The rest of the standard connections are left as homework assignments.

Most three-phase connections can be solved by considering total kVA and the $\sqrt{3}$ relation for phase and line quantities, as is done in Example 6.2. However, it is probably more instructional to begin with the more detailed presentation in Example 6.1.

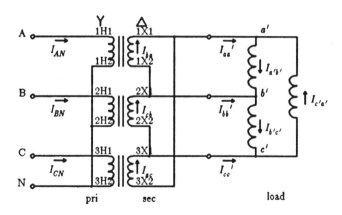

Figure 6.1 Wye-delta transformer connection showing polarity markings.

Example 6.1

Wye-delta transformer connection. A balanced load of 800 kW, 480 V, and an 80 percent lagging power factor is to be supplied from a 12,470 V substation through a wye-delta transformer bank. Determine all voltages and currents, and then sketch the complete phasor diagrams. Neglect exciting current and impedance voltage drops. Assume an *abc* sequence with $V_{AB} = 12,470\underline{/0^\circ}$ V as shown in Figure 6.2.

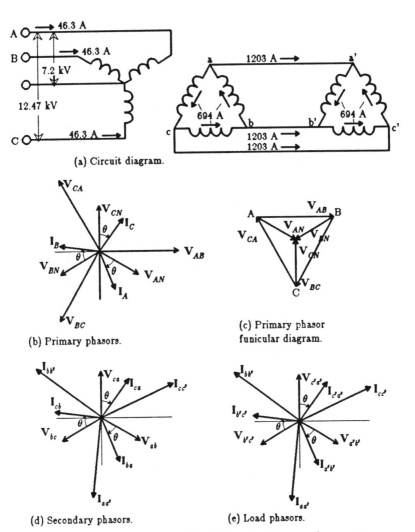

(a) Circuit diagram.

(b) Primary phasors.

(c) Primary phasor
funicular diagram.

(d) Secondary phasors.

(e) Load phasors.

Figure 6.2 Wye-delta connection of Example 6.1 for an *abc* sequence.

Solution

Step 1: Determine the primary voltages

$$V_{H\ line} = 12.47 \text{ kV with } V_{AB} = 12.47 \underline{/0^\circ} \text{ kV}$$

$$V_{H\ phase} = \frac{12.47}{\sqrt{3}} = 7.20 \text{ kV}$$

This system is usually referred to by the electric utility industry as a 7.2Y/12.5 kV system. Since $V_{AB} = 12.47\underline{/0^\circ}$ kV, the other line voltages are displaced 120° with $V_{BC} = 12.47\underline{/-120^\circ}$ kV, and $V_{CA} = 12.47\underline{/120^\circ}$ kV. The funicular diagram of Figure 6.2c is formed by plotting V_{AB} at $\underline{/0^\circ}$. Then from the end of the V_{AB} phasor, extend the V_{BC} phasors in the direction of $1.0\underline{/-120^\circ}$. Finally, extend the phasor V_{CA} from the end of V_{BC} in the direction of $1.0\underline{/120^\circ}$. Note that the phase voltages for the primary transformer coils are displaced 30° from the line voltages and extend from the corners of the triangle as $V_{AN} = 7.2\underline{/-30^\circ}$ kV, $V_{BN} = 7.2\underline{/-150^\circ}$ kV, and $V_{CN} = 7.2\underline{/90^\circ}$ kV. For clarification, replot the phase and line phasor voltages as shown in Figure 6.2b.

Step 2: Determine the secondary voltages. Since the load was given as delta-connected at 480 V

$$V_{X\ line} = V_{X\ phase} = 480 \text{ V}$$

The phase angle of the ideal transformer primary and secondary voltages are in phase with each other; therefore, the secondary voltages are:

$$V_{ab} = 480\underline{/-30^\circ}, \quad V_{bc} = 480\underline{/150^\circ}, \quad \text{and} \quad V_{ca} = 480\underline{/90^\circ}.$$

as shown in the phasor diagram, Figure 6.2d. For additional clarity, the load voltages, which are identical to the transformer secondary voltages, are plotted in Figure 6.2e.

Step 3: Determine the transformer coil turns ratio.

$$\frac{V_{H\ phase}}{V_{X\ phase}} = \frac{7.20}{0.480} = 15.0 = a = \text{turns ratio}$$

Step 4: Compute the load currents from the expression

$$P_{load} = \sqrt{3}V_X I_X \cos\theta$$

$$800{,}000 = \sqrt{3}(480)I_X (0.8)$$

$$I_X = I_{X\ line} = 1203 \text{ A}$$

$$I_{X\ phase} = \frac{I_{X\ line}}{\sqrt{3}} = 694 \text{ A}$$

The phase $a'b'$ load current, $I_{a'b'}$, lags behind the phase voltage, $V_{a'b'}$, by a power factor angle of 36.8° (pf = 0.8). Thus, $I_{a'b'} = 694\underline{/-66.9^\circ}$. The other phase currents are rotated $\pm120^\circ$. The line currents are $\sqrt{3}$ larger and lag the phase currents by 30° as shown in Figure 6.2e.

Step 5: Compute the transformer secondary currents. The simplest method for a delta load with a delta-connected secondary is to note that the transformer secondary currents are in phase with the load currents. Thus, $I_{ba} = I_{a'b'} = 694\underline{/-66.0°}$. The other currents are offset by 120°.

Step 6: Compute the primary currents by using the following equations:

$$\frac{I_{H\ phase}}{I_{X\ phase}} = \frac{1}{a}$$

$$I_{H\ phase} = \frac{I_{X\ phase}}{a} = \frac{694}{15.0} = 46.3\ \text{A}$$

$$I_{H\ line} = I_{H\ phase} = 46.3\ \text{A}$$

Assuming ideal transformers, the transformer primary currents are in phase with the load currents. Thus,

$$I_{AN} = \frac{I_{ba}}{a} = \frac{I_{a'b'}}{a} = \left(\frac{694}{15.0}\right)\underline{/-66.9°} = 46.3\underline{/-66.9°}\ \text{A}$$

Note that the primary currents can be found more directly by considering the source power, where

$$P_{source} = \sqrt{3}V_H I_H \cos\theta$$

$$I_{H\ line} = \frac{800}{\sqrt{3}(12.47)(0.8)} = 46.3\ \text{A}$$

Check: It is a good practice to verify the results. Remember that the source kVA and power should equal the load kVA and power, thus

$$\text{kVA}_{source} = \sqrt{3}V_H I_H = \sqrt{3}(12.47)(46.3) = 100\ \text{kV}$$

$$P_{source} = S\ \cos\theta = 1000(0.80) = 800\ \text{kW}$$

These values agree with the given data.

Example 6.2

Repeat Example 6.1 for a delta-wye transformer connection with the same source and load. Use the simplified procedure by considering only the magnitude of the currents and voltages.

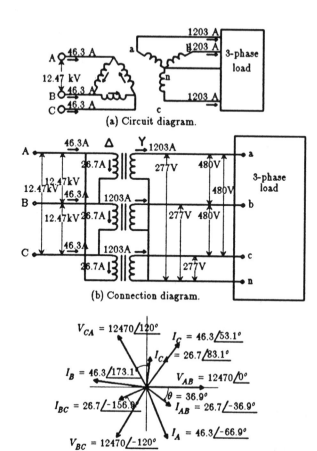

(a) Circuit diagram.

(b) Connection diagram.

(c) Primary phasors.

(d) Secondary phasors.

Figure 6.3 Delta-wye connection of Example 6.1 for an *abc* sequence.

Solution

The secondary line current is found from $S = \sqrt{3} V_L I_L$ as

$$I_L = \frac{1,000,000}{\sqrt{3}(480)} = 1203 \ A$$

The secondary line-to-neutral is $\frac{480}{\sqrt{3}}$ or 277 V. The primary line voltage is the same as the transformer coil voltage of 12,470 V. This means that each transformer turns ratio is

$$a = \frac{12,470}{277} = 45.0$$

The primary coil current is then $1203/a = 26.7$ A and the primary line current is $\sqrt{3}(26.7) = 46.3$ A. As a check on arithmetic consider that the secondary coil kVA is

$$V_{coil} \ I_{coil} = 277 \times 1203 = 333 \ kVA$$

which is one-third of the load kVA. The primary coil kVA is $12,470(26.7) = 333$ kVA, which is also one-third of the load kVA. The primary line current could also be calculated from the equation $S = \sqrt{3} V_L I_L$ so that

$$I_L = \frac{1,000,000}{\sqrt{3} \times 12,470} = 46.3 \ A$$

Then the transformer coil current is $46.3/\sqrt{3} = 26.7$ A. The above results are shown in Figure 6.3.

Example 6.3

Compare the four standard types of three-phase connections for the source and load of Example 6.1 where the source is 12,470 V and three-phase, and the load is 480 V and 1000 kVA with an 80 percent power factor.

Solution

The secondary line current is

$$I = \frac{S}{\sqrt{3} V} = \frac{1000 \times 10^3}{\sqrt{3} \times 480} = 1203 \ A$$

Connection	Primary				Secondary				Turns Ratio	kVA Per Phase	kVA Three Phase
	Line		Coil		Line		Coil				
	Volts	Amps	Volts	Amps	Volts	Amps	Volts	Amps			
(a) Delta-wye	12,470	46.3	12,470	26.7	480	1203	277	1203	45	333	1000
(b) Wye-delta	12,470	46.3	7,200	46.3	480	1203	480	694	15	333	1000
(c) Delta-delta	12,470	46.3	12,470	26.7	480	1203	480	694	26	333	1000
(d) Wye-wye	12,470	46.3	7,200	46.3	480	1203	277	1203	26	333	1000

Figure 6.4 Tabulated results of Example 6.3.

The primary line current is

$$I = \frac{1000 \times 10^3}{\sqrt{3} \times 12{,}470} = 46.3 \;\; A$$

A summary of the various connections is given in Figure 6.4.

(a) *Delta-wye connection.* The secondary line voltage is 480 V. The transformer line-to-neutral coil voltage is

$$V_{2\;phase} = V_{2\;coil} = \frac{480}{\sqrt{3}} = 277 \;\; V$$

The secondary coil current is the line current of

$$I_{2\;coil} = I_{2\;line} = 1203 \;\; A$$

The primary coil voltage is the same as the line-to-line voltage for the delta connection of 12,470 V. The primary coil current for the delta connection is $I_{line}/\sqrt{3} = 46.3/\sqrt{3} = 26.7$ A. A second method for determining the current is to consider the turns ratio of each transformer coil which is $12{,}470/277 = a = 45.0$. The $I_{pri\;coil} = I_{sec\;coil}/a = 1203/45.0 = 26.7$ A.

(b) *Wye-delta connection.* The secondary line-to-line and phase voltages of 480 V are identical. The secondary line current is

$$I_{2\;line} = \frac{1000 \times 10^3}{\sqrt{3} \times 480} = 1203 \;\; A$$

The transformer secondary coil current is $1203/\sqrt{3} = 694$ A. The transformer primary coil voltage is $1/\sqrt{3}$ times the source line-to-line voltage.

$$V_{1\;coil} = \frac{12{,}470}{\sqrt{3}} = 7200 \;\; V$$

The primary coil current is the same as the source line current of 46.3 A. A second method for determining the current ratio is to begin with the transformer coil turns ratio of a 7200/480 = 15. Then the primary coil current is

$$I_{1\ coil} = \frac{I_{2\ coil}}{a} = \frac{694}{15} = 46.3\ \text{A}$$

(c) *Delta-delta connection.* The primary and secondary coil voltages are the same as the line-to-line voltages of 12,470 V and 480 V, respectively. The secondary coil current is $1203/\sqrt{3} = 694$ A and the primary current is $46.3/\sqrt{3} = 26.7$ A. The coil turns ratio is 12,470/480 = 26.0.

(d) *Wye-wye connection.* The primary coil voltage is $12,470/\sqrt{3} = 7200$ V. The secondary coil voltage is $480/\sqrt{3} = 277$ V. The primary and secondary coil currents are the same as the line currents of 46.3 V and 1203 A, respectively. The coil turns ratio is 7200/277 = 26.0.

A summary of this example is shown as Figure 6.4.

(a) The ideal shape of a three-phase transformer.   (b) Three-phase core-type transformer.   (c) Three-phase shell-type transformer.

Figure 6.5 Three-phase transformer construction.

B. Polyphase Transformer Construction

The polyphase transformer configurations may be obtained by using three single-phase transformers or by using a specially constructed three-phase transformer, as shown in Figure 6.5. A three-phase transformer consists of three sets of primary and secondary windings wound on a specially constructed combination core. Consider three single-phase transformers connected for transforming three-phase power. The transformers could be designed with the primary and the secondary windings both wound on the same leg of each respective core. If three free sides of the cores are butted together, as shown in Figure 6.5a, the fluxes through the butted sides will be equal in magnitude and 120° out of phase. Therefore, the resultant flux through the three combined butted cores will be zero. The iron of the three butted sides, therefore, could be removed without affecting the operation of the transformers. This relation results in the construction utilized for three-phase transformers. The general features of construction are shown in Figure 6.5b for core-type construction and in Figure 6.5c for shell-type construction. This means that there is less core material, fewer terminal insulators, less insulating oil, and a smaller case or container. Thus, a three-phase transformer is less expensive than three single-phase transformers.

The windings of three-phase transformers may be wye- or delta-connected in the same manner as for three single-phase transformers. The interconnections are often made inside the case so that only the line terminations need to be brought outside the case, as shown in Figure 6.6.

A three-phase transformer is rated in terms of its three-phase rating. For example, the system of Example 6.1 could be *one* 1000 kVA, 12,470:480 V, wye-delta, three-phase transformer, or it could be *three* 333 kVA, 7200:480 V, single-phase transformers connected wye-delta. The voltage and current ratings of the transformer coils are those given in Example 6.1.

C. Advantages and Disadvantages of Various Three-phase Transformer Connections

1. *Wye-wye*

   a. Coil voltage is $1/\sqrt{3}$ or 57.7 percent of line voltage.
   b. Coil current equals line current with larger conductor cross section. This generally means greater mechanical strength.
   c. Lower voltage means a lower flux level and smaller core size.
   d. The lower voltage (of a) and larger cross section (of b) and the smaller size (of c) normally mean a less expensive transformer.
   e. Secondary and primary voltages are approximately in phase.
   f. The wye-wye connection with neutrals tied together reduces the chance of the ferroresonance problem.
   g. A third harmonic current shows up on the neutral line, which causes interference on telephone signals.
   h. If the neutral is not connected an unbalanced load will cause unbalanced voltages.

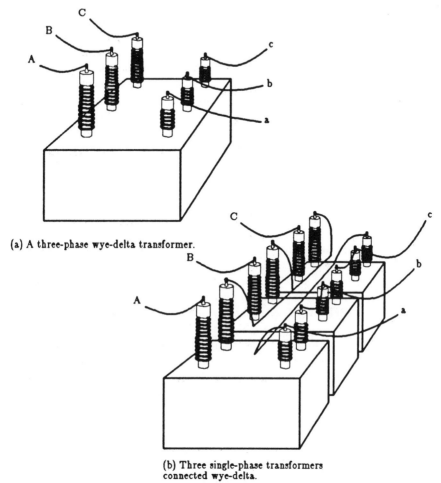

(a) A three-phase wye-delta transformer.

(b) Three single-phase transformers connected wye-delta.

Figure 6.6 Three-phase transformer systems.

2. *Wye-wye with Delta Tertiary (also called Wye-delta-wye)*

When it is necessary to have a wye-wye connection with a high impedance neutral or no neutral, as with long transmission systems, a third winding called a *tertiary* winding is added. The tertiary winding is delta-connected and eliminates the problems of the unbalanced voltages and third harmonic currents. The tertiary winding may be unloaded or it may be used to supply voltages for lights, pumps, fans, or loads for small communities near a substation.

3. *Delta-delta*
    a. Balanced voltages for full load range of operation.
    b. Third harmonic currents circulate in closed delta and do not show up in line voltages.
    c. If one transformer breaks down or is removed, the system will operate at reduced load as an open-delta connection.
    d. Secondary voltages are approximately in phase with the primary voltages.
    e. The transformers must have the same turns ratio and equivalent impedance; otherwise, there will be a circulating current in the transformer even when there is no connected load.
    f. Grounding of the system is more difficult to accomplish. (See Section 6.3 on four-wire delta connections.)

4. *Wye-delta and Delta-wye*
    a. Balanced voltages for full load range.
    b. There is a 30° phase shift between the primary and secondary voltages. (See Figures 6.2 and 6.3.)
    c. Usually used with the wye coils on the high voltage side and delta coils on the low voltage side.
    d. Grounding on the secondary of the wye-delta connection is more difficult to accomplish. (See Section 6.3 on the four-wire delta connection.)
    e. There are possible problems with ferroresonance when the load is supplied through cables.

D. Effects of Reversed Polarity

The polarity of each transformer of a three-phase bank of transformers must be properly connected. If not, improper voltages result for a wye secondary, and a short circuit is created for a delta secondary. First, consider details of the circuit with the wye secondary. Consider the case of the delta-wye connection of Figure 6.7 with the secondary coil of the transformer of phase c reversed. The primary voltages are supplied from a balanced source and are shown in Figure 6.7b. The phasor of V_{cn} is 180° from V_{CA} as shown in Figure 6.7c. Then the line-to-line voltages are shown in Figure 6.7e. Thus, it is noted that two of the line voltages have the same magnitude as the phase voltage, and one line voltage is $\sqrt{3}$ times larger than the phase voltage. This is further emphasized in Example 6.4.

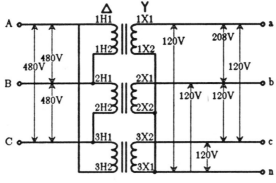

(a) Connection diagram.

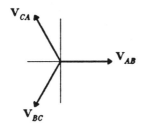

(b) Delta primary voltages.

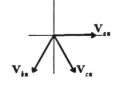

(c) Wye secondary voltages.

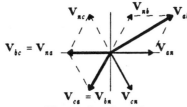

(d) Secondary line voltages.

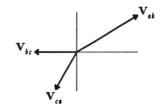

(e) Line-to-line secondary voltages.

Figure 6.7 Delta-wye connection with one transformer secondary coil of reversed polarity.

Example 6.4

Reversed polarity for a wye-connected secondary. Consider the delta-wye circuit of Figure 6.7 for three 480:120 V transformers. Determine the secondary line and phase voltages with the secondary of the CA-cn transformer coil reversed.

Solution

Sketch the primary and secondary voltage phasors as shown in Figure 6.7b and c. Then add the secondary phase voltage as shown in Figure 6.7d to obtain the line voltages as shown in Figure 6.7e. Note that the phase voltages are all 120 V, that two of the line voltages are also 120 V, and that only one line voltage is 208 V. Also, note that the line voltages are not separated by $\pm 120°$.

The effect of a reversed coil on a delta secondary system is more difficult to prove and the result is more traumatic than for the wye secondary. With a reversed coil in the system, a large current will circulate in the secondary delta loop. This current causes large primary currents to flow. The result could be burned or damaged coils or, at best, tripped fuses or circuit breakers. This result is explained in Example 6.5 and Figure 6.8.

The reversal for either the wye or delta secondary system could be accomplished by reversing one or several coils on either the secondary or primary side. The examples have shown the reversal on the secondary only. It is true that if all coils are reversed there is no problem.

Example 6.5

Reversed polarity for a delta-connected secondary. Determine the voltage between terminals f and g of Figure 6.8 for the delta secondary with the secondary of transformer $\overline{CA}$ and $\overline{ca}$ reversed.

Solution

The delta circuit is left open as shown in Figure 6.8a. The voltage difference between terminals f and g is

$$V_{gf} = V_{fa} + V_{ab} + V_{bc} + V_{cg}$$
$$= 0 + 240\underline{/0°} + 240\underline{/-120°} + 240\underline{/-60°}$$
$$= 480\underline{/-60°} \text{ V}$$

The voltage is twice the line-to-line magnitude. If terminals f and g are tied together, a very large current will flow in the delta loop since only the low impedance of the coil limits the amount of current flow. If the large currents are not interrupted by appropriate fuses or circuit breakers, the transformers could be damaged.

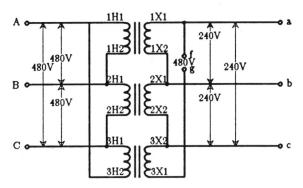

(a) Connection diagram.

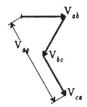

(b) Primary voltages.

(c) Secondary voltages.

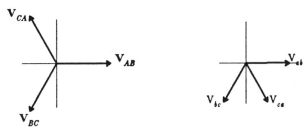

(d) Sum of secondary voltage with *fg* open.

Figure 6.8 Delta-delta connection with phase $\overline{CA}$ transformer coil of reversed polarity.

6.2 OPEN-DELTA CONNECTION

Two single-phase transformers may be connected in what is called "open-delta" or "V-V" as shown in Figure 6.9 to supply a three-phase load from a three-phase supply. There are three common applications for this connection: (1) for emergency use of a delta-delta system with one transformer

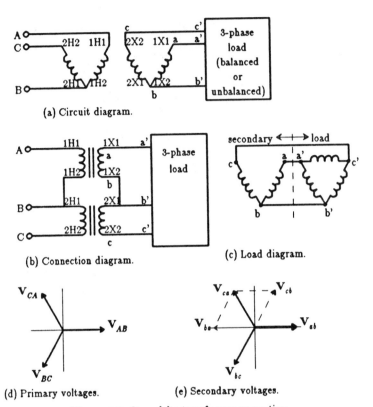

(a) Circuit diagram.

(b) Connection diagram.

(c) Load diagram.

(d) Primary voltages.

(e) Secondary voltages.

Figure 6.9 Open-delta transformer connection.

out of service, (2) for the initial phase in the operation of an industrial plant, (3) where the two transformers are of different size and one transformer supplies a single phase load while the two transformers combined supply a three-phase system.

The theory of operation is depicted in Figure 6.9d. Note that balanced three-phase voltages are supplied to the primary. These voltages produce an excitation of I_{AB} in transformer 1 and I_{BC} in transformer 2. These currents cause magnetic fluxes that induce the currents I_{ba} and I_{cb} in the secondaries of the respective transformers with the related voltages V_{ab} and V_{bc}. The voltage measured between terminals c and a is the sum V_{cb} and V_{ba}, which is vectorially the balanced value of V_{ca}. Hence, the secondary of the open delta provides balanced three-phase voltages. These three-phase voltages remain with balanced or unbalanced load currents. An unbalanced case is shown in Figure 6.10.

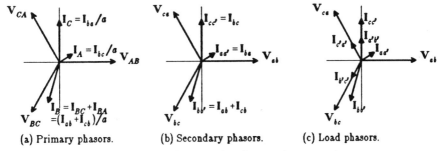

(a) Primary phasors.   (b) Secondary phasors.   (c) Load phasors.

Figure 6.10 Open-delta circuit voltage and current phasors for an unbalanced load.

The disadvantage of the open-delta connection is that the transformers must have a kVA rating larger than the load kVA, as noted in (6.3) and (6.6). In order to explain this increased rating, consider that three single-phase transformers connected in delta-delta must supply

$$\text{kVA}_{delta\text{-}delta} = 3V_{TR}I_{TR} = \text{total kVA} \tag{6.1}$$

where V_{TR} and I_{TR} are the rated values of the voltage and current coils of the transformers. The two transformers in open-delta must supply a three-phase load and can be described as

$$\text{kVA}_{vv} = \sqrt{3}V_{TR}I_{TR} = \text{total kVA} \tag{6.2}$$

The ratio of (6.2) to (6.1) is

$$\frac{\text{kVA}_{vv}}{\text{kVA}_{delta\text{-}delta}} = \frac{\sqrt{3}V_{TR}I_{TR}}{3V_{TR}I_{TR}}$$

$$= \frac{\sqrt{3}}{3} = 0.577 \text{ or } 57.7\% \tag{6.3}$$

Thus (6.3) relates the kVA rating of two transformers in open-delta to that of three transformers in delta-delta.

Another important consideration is the rating of each of the transformers in the open-delta circuit. Each of the open-delta transformers is rated $\text{kVA}_{vv'}$, which is half the total kVA load, or

$$\text{kVA}_{vv'} = \frac{1}{2}\left(\sqrt{3}V_{TR}I_{TR}\right) \tag{6.4}$$

Since the rating of each transformer in a delta-delta connection is

$$kVA_{delta-delta'} = V_{TR}I_{TR} \tag{6.5}$$

the ratio of (6.4) to (6.5) is

$$\frac{kVA_{vv'}}{kVA_{delta-delta'}} = \frac{(\sqrt{3}/2)V_{TR}I_{TR}}{V_{TR}I_{TR}} \tag{6.6}$$

$$= \frac{\sqrt{3}}{2} = 0.866 \quad \text{or} \quad 86.6\%$$

Example 6.6

Three 100 kVA transformers connected delta-delta are supplying a 300 kVA load. (a) What load could be supplied if one of the transformers is removed to form an open-delta circuit? (b) What load is supplied by each of the open-delta transformers of part (a)?

Solution

Three 100 kVA transformers will supply a 300 kVA load. Two transformers in open-delta will supply

$$kVA_{vv} = 0.577 \times 300 = 173 \text{ kVA}$$

Each of the transformers must then supply half of the 173 kVA

$$kVA_{vv'} = \frac{173}{2} = 86.6 \text{ kVA}$$

Note that each 100 kVA rated transformer connected in open-delta provide only 86.6 kVA or 86.6% of its rating.

Example 6.7

An open-delta bank of transformers supplies a 300 kVA load. What is the required kVA rating of each transformer?

Solution

Each transformer of the open-delta bank must supply half the total load of $300/2 = 150$ kVA. According to (6.6), each transformer is operated at 86.6 percent of rated capacity. Therefore, each transformer must be rated at 150/0.866 = 173 kVA. This means that three 173 kVA transformers in closed-delta will supply a 520 kVA load or two 173 kVA transformers in open-delta will supply 520(0.577) = 300 kVA of load.

6.3 WYE AND DELTA FOUR-WIRE CIRCUITS

Many modern buildings now use the 120/208 three-phase system for power distribution, is shown in Figure 6.11. This system provides the advantage of three-phase while retaining the capability of providing for standard single-phase 120 V equipment.

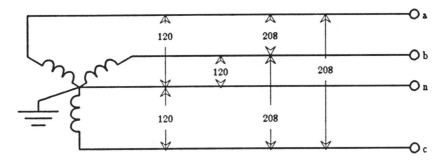

Figure 6.11 Four-wire wye 120/208 V.

Another common system is the four-wire delta circuit shown in Figure 6.12b. This is often used when three-phase equipment is added to an existing single-phase system since all previous installations can remain undisturbed. The major disadvantage to the four-wire delta system is the 208 V line-to-neutral voltage on what is called the "high-leg," "red-leg," "sting-leg," or "wild-leg." The National Electrical Code (Section 215.8) requires that this circuit be identified by an outer finish that is orange in color or by tagging or other effective means. This high leg is normally phase B.

6.4 SPECIAL CONNECTIONS OF THREE-PHASE TRANSFORMERS

The primary and secondary line voltages for wye-wye and delta-delta connections are in phase and produce no special problems. However, there is a 30° phase shift between the primary and secondary line voltages for the

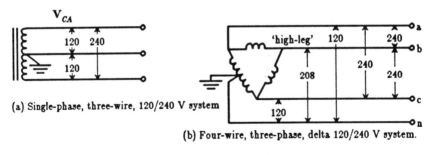

(a) Single-phase, three-wire, 120/240 V system

(b) Four-wire, three-phase, delta 120/240 V system.

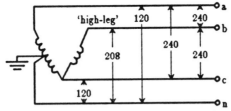

(c) Four-wire, three-phase, open-delta, 120/240 V system.

Figure 6.12 Several 120/240 V systems.

wye-delta connections. The immediately obvious connection places the secondary line voltage for the wye-delta, is shown in Figure 6.13, and vice versa for the delta-wye connection. If the connections are made in a less obvious manner, the two connections can be made to agree. Therefore, the electrical industry through ANSI has set the standard that the connections shall be made so that the high potential line voltages shall lead the low potential line voltages by $30°$. The following example should suffice in explaining this procedure.

Example 6.8

Two transformer banks are to be connected in parallel for supplying 480 V from a 4160 V system. One transformer bank consists of three 4160:277 V transformers to be connected delta-wye, and the other bank consists of three 2400:480 V transformers connected in wye-delta. Show the correct ANSI connection.

Solution

The high side line voltages should lead the low side line voltages by $30°$, as is shown in Figure 6.13.

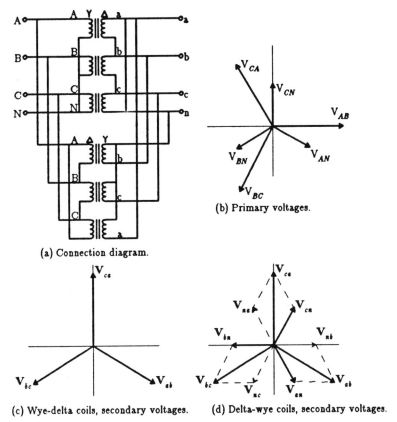

(a) Connection diagram.

(b) Primary voltages.

(c) Wye-delta coils, secondary voltages.

(d) Delta-wye coils, secondary voltages.

Figure 6.13 Parallel connection of delta-wye and wye-delta transformers.

PROBLEMS

6.1 Sketch the connection diagram and circuit diagram similar to Figure 6.3a and 6.3b for the (a) wye-delta, (b) delta-delta, and (c) wye-wye connections.

6.2 The high voltage terminals of a three-phase bank of three single-phase transformers are connected to a three-wire, three-phase, 13,800 V (line-to-line) system. The low voltage terminals are connected to a three-wire, three-phase substation load rated at 1500 kVA and 4160 V line-to-line. Specify the voltage, current, and kVA ratings of each transformer (both high and low voltage windings) for the following connections: (a) delta-wye, (b) delta-delta, (c) wye-wye, and (d) wye-delta.

6.3 Three 37.5 kVA, 4800:120 V, single-phase transformers are connected wye-wye step down. (a) Show the connection or circuit diagram. (b) What are the primary and secondary line voltages? (c) What are the rated line currents for the supply and load sides? (d) What is the total kVA for rated load?

6.4 Given three single-phase transformers connected delta-wye and each transformer rated 75 kVA, 4160:120 V, determine all the rated voltages and currents and sketch the phasor diagrams for the primary and secondary quantities in a manner similar to that of Example 6.2 (assume rated load current at 0.9 lagging power factor).

6.5 Given one delta-wye, three-phase transformer rated 225 kVA, 4160:208 V, sketch the rated voltage and current phasor diagrams using the simplified procedure of Example 6.2 (assume rated load current at 0.9 lagging power factor).

6.6 Given three single-phase transformers connected wye-delta, and each transformer rated 167 kVA, 7200:480 V, determine all the rated voltages and currents and sketch the phasor diagrams for the primary and secondary quantities in a manner similar to that of Example 6.2. Assume the rated load current has a 0.95 lagging power factor.

6.7 Given one wye-delta, three-phase transformer rated 500 kVA, 12,470:480 V sketch the rated voltage and current phasor diagrams using the simplified procedure of Example 6.2. Assume the rated load current has a 0.95 lagging power factor.

6.8 Mr. Allen is having a new grocery store constructed. It has been determined that the load will be 600 A at 480 V, three-phase, four-wire. Provo City Power will provide three pole-mounted single-phase transformers connected delta-wye and supplied from a 12,470 V, three-phase source. (a) What is the required kVA rating of each transformer? (b) What is the approximate current in the primary mains?

6.9 A three-phase, wye-delta transformer is used to take power from a three-phase supply at a line voltage of 13,200 V and deliver to a three-phase load at a line voltage of 480 V. The total load is 2500 kVA. At rated load, determine (a) the high-side and low-side line currents, (b) the transformer coil voltages, (c) the transformer coil currents, (d) the kVA rating of the transformer, and (e) how the line-voltage ratio compares with the transformer turns ratio.

6.10 The high-voltage coils of a 112.5 kVA, 7200:208 V, three-phase, delta-wye transformer is connected to a 7200 V source. The low voltage coils are connected to a balanced-wye resistive load with 0.3 Ω per phase.

(a) Sketch the circuit. (b) Calculate the secondary line current. (c) Calculate the approximate source line current.

6.11 Three 1667 kVA, 7200:2400 V single-phase transformers are connected three phase at a small generating station to step up the output of a 2400 V three-phase generator to a 12,470 V transmission system. For the rated load, determine the (a) total kVA transmitted over the power line, (b) line voltages and currents on the primary and secondary side of the transformer, and (c) ratio of line voltages.

6.12 A three-phase motor requires 60 kW at an 84 percent lagging power factor when operated at 208 V. The motor is supplied from a three-phase 4160:208 V, wye-delta, step-down transformer bank. Find the line currents on the (a) high-voltage and (b) low-voltage sides of the transformer.

6.13 Three 2400:120 V, single-phase transformers are connected in delta-wye. What are the approximate secondary terminal voltages if the polarity of the primary coil of the transformer in leg BC-bn is reversed?

6.14 Three 2400:120 V, single-phase transformers are connected in wye-wye. What are the approximate secondary terminal voltages if the polarity of the secondary coil of the transformer in leg CN-cn is reversed?

6.15 Three 2400:277 V, single-phase transformers are connected in delta-wye. What are the approximate secondary terminal voltages if the polarity of the primary coil in leg BC-bn is reversed?

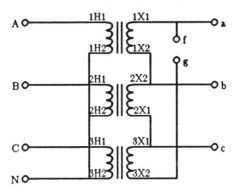

Figure 6.14 Circuit for Problem 6.16.

6.16 Three 277:240 V single-phase transformers are connected in wye-delta to a 480 V three-phase source shown in Figure 6.14. (a) What is the voltage between terminals *f* and *g* if coil *bc* is of reversed polarity? (b) What would happen if terminals *f* and *g* were connected together?

6.17 An electrician reports that the open circuit line-to-line voltages on the secondary side of a delta-wye connection are 120 V, 120 V, and 208 V. What is a possible problem with the system?

6.18 An electrician reports that the fuses in the primary source blow every time the switch is closed on the source side of a set of three single-phase transformers connected in wye-delta. There was no load connected to the secondary, and the transformers tested satisfactorily in a single-phase connection. What is a possible reason for the trouble?

6.19 A 1500 kVA, 2400 V three-phase load is to be supplied from two single-phase transformers connected in open-delta. What is the required kVA rating of each of the single-phase transformers?

6.20 Two 37.5 kVA, 2400:240 V, single-phase transformers are connected open-delta. Calculate the three-phase, wye-connected, resistive load (a) in kVA and (b) in Ω per phase that can be connected to the secondary terminals without overrating the transformer.

6.21 Three 333 kVA, 4160:240/120 V, 60 Hz, single-phase distribution transformers are connected delta-delta in order to provide a 240 V four-wire delta service. (a) Sketch the circuit. (b) Calculate the secondary voltage V_{ab}, V_{ac}, V_{an}, etc. (c) What problems would arise if the secondary of this system was connected in parallel with a four-wire, wye-wye system?

6.22 Given a 225 kVA, 12,470Y/7200:208Y/120 V, wye-wye, three-phase transformer with the neutrals of the primary and secondary coils solidly grounded and a load connected between a and n, determine the approximate voltage across the load.

6.23 Repeat Problem 6.21 with the neutral connection deleted and the load connected between terminals a and n.

6.24 Repeat Example 6.5 with the primary coil of transformer AB-ab of reversed polarity. What will happen when the terminals f and g are connected together?

6.25 Given a 150 kVA, wye-delta, 12,470Y/7200:240 V, three-phase transformer, a single-phase load is connected between lines a and b at the secondary of the transformer. What is the approximate voltage across the load?

6.26 Three 50 kVA, 480:120 V single-phase transformers are connected as shown in Figure 6.15. The line-to-line source voltage is of such a magnitude that for normal connection, the transformer coil would be operated at rated voltage. What are the approximate magnitudes of the secondary line-to-neutral voltages (V_{an}, V_{bn}, V_{cn}) and the secondary line-to-line voltages (V_{ab}, V_{bc}, V_{ca}) for the transformer connections of (a) Figure

6.15a, (b) Figure 6.15b, (c) Figure 6.15c, (d) Figure 6.15d, (e) Figure 6.15e, and (f) Figure 6.15f.

	V_{ab}	V_{bc}	V_{ca}	V_{an}	V_{bn}	V_{cn}	Remarks
a							
b							
c							
d							
e							
f							

6.27   Three 2400:277 V, single-phase transformers are connected in wye-delta. What are the secondary line and phase voltages if the polarity of one of the transformer coils is reserved?

6.28   Three 50 kVA, 2400:208 V, single-phase transformers are connected wye-delta step-down as per Figure 6.6b. (a) Show the connection diagram. (b) What are the primary and secondary line voltages? (c) What are the currents for the source and load feeder lines for rated transformer currents? (d) What is the load kVA for rated transformer currents?

6.29   A 150 kVA, 4160:208 V, three-phase transformer is connected wye-delta step-down as per Figure 6.6a. (a) Show the connection diagram. (b) What are the primary and secondary line voltages? (c) What are the currents for the source and load feeder lines for rated transformer currents? (d) What is the load kVA for rated transformer currents?

6.30   Three single-phase, pole-mounted transformers connected wye-wye supply a department store. Each transformer is rated 100 kVA, 7200:120 V, 60 Hz and has a series impedance of (5.8 + j8.9) Ω referred to the high-voltage coil. Determine the following: (a) rated secondary line current, (b) rated secondary line-to-line voltage, (c) rated primary line current, (d) rated primary line-to-line voltage, and (e) primary and secondary line current for a three-phase short-circuited fault at the transformer secondary terminals.

6.31   A three-phase, 300 kVA, 12,470:208 V, 60 Hz, wye-wye pad-mounted transformer is installed to supply a department store. The series impedance is (5.8 + j8.9) Ω and is referred to the high-voltage coil. Determine the following: (a) rated secondary line current, (b) rated secondary line-to-line voltage, (c) rated primary line current, (d) rated primary line-to-line voltage, (e) primary and secondary line current for a three-phase short-circuited fault at the transformer secondary terminals.

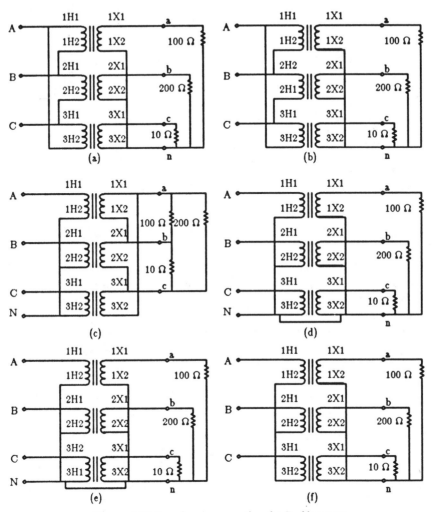

Figure 6.15 Transformer connections for Problem 6.26.

6.32 The 25 kW of 277 V fluorescent lights in a modern variety drug-store are supplied from a 480/277 V wye system. Assume a 0.95 lagging power factor. The 480/277 V supply is obtained from a three-phase wye-wye transformer bank with a primary line voltage of 2400 V. (a) What is the transformer primary and secondary current rating? (b) What size of copper conductor with THW insulation is required for the transformer primary and secondary feeders? (c) What size fuse would you recommend for the primary and secondary side of the transformer?

6.33 An industrial plant has a 2770 V three-phase distribution system. An 800 kVA, three-phase rectifier system requires a 2400 V three-phase source. It has been decided to use a three-phase, step-down, wye-wye, autotransformer to handle this difficulty. (a) Sketch the circuit. (b) What is the input line current to the rectifier? (c) What is the line current supplied by the 2770 V system? (d) What is the required current for each coil of the autotransformer?

7

POLYPHASE INDUCTION MOTORS

7.1 INTRODUCTION

There are more induction motors (Figure 7.1) in regular use than any other type of electric motor. The principle of operation requires a rotating magnetic field as described in Chapter 4. This rotating field, which is provided by the stator windings, is the same as that required for a synchronous machine.

The squirrel-cage rotor is a set of conductors attached to two end rings much like a hamster, gerbil, or squirrel cage. Electric currents are induced in these rotor conductors by transformer action from the alternating currents flowing in the stator windings. If the rotor windings are insulated and connected to slip rings at one end of the shaft, the assembly is referred to as a wound-rotor induction motor.

Although the induction machine is normally used as a motor, it can also be used as a generator. When used as a generator, it is called an induction generator or asynchronous generator.

How does an induction motor operate? To explain the principle of operation, consider a closed loop conductor fixed in space with a set of magnetic poles rotating counterclockwise as shown in Figure 7.2. A voltage is induced in the conductor according to (7.1).

$$\vec{e} = l \, (\vec{u} \times \vec{B}) \tag{7.1}$$

where $\vec{e}$ is the voltage vector

l is the length of the conductor in meters

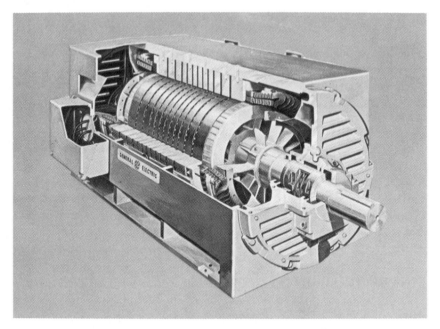

Figure 7.1 A polyphase squirrel-cage induction motor. *(Courtesy of General Electric Company)*

$\vec{u}$ is the velocity of the conductor in m/s

$\vec{B}$ is the flux density of the magnetic field in teslas

For the example of Figure 7.2a, the flux density vector directed from the north pole to the south pole and the motion of the upper conductor with respect to the magnetic field is clockwise. Thus, the vectors of (7.1) indicate a voltage vector into the page, as shown in Figure 7.2b. The voltage will cause a current to flow into the page for the upper conductor of Figure 7.2a. According to (7.2), a force causes the upper conductor to move in a counter-clockwise direction as a result of the current and flux, as shown in Figure 7.2c. The magnetic force equation is

$$\vec{F} = i\vec{l} \times \vec{B} \tag{7.2}$$

where $\vec{F}$ is the force in newtons

$\vec{i}$ is the electric current in amperes

l is the length of the conductor in meters

$\vec{B}$ is the flux density in teslas

Since there must be motion between the rotating magnetic field and the conductor to cause the induced current, the rotor coil will rotate at a speed

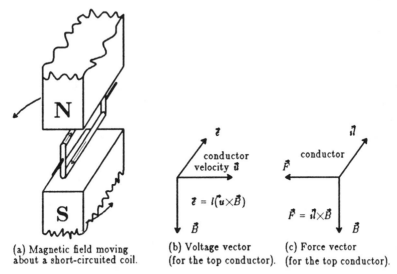

(a) Magnetic field moving about a short-circuited coil.

(b) Voltage vector (for the top conductor).

(c) Force vector (for the top conductor).

Figure 7.2 Induction motor theory.

slower than the speed of the rotating field. This difference in speed is called the slip speed.

For the practical induction motor, the magnetic field is created by the multiple phase windings in the same manner as for the synchronous motor described in Chapter 4. The rotor contains numerous conductors. Current is induced by transformer action in the rotor conductors when the rotating stator field moves at a speed different than the rotor conductors. This difference is expressed as a ratio of the difference in the rotor speed from the stator field speed with respect to the stator field speed and is called slip. It is defined as

$$s = \frac{n_s - n_r}{n_s} = \frac{\omega_s - \omega_r}{\omega_s} \tag{7.3}$$

where s is the per unit slip

n_s is the synchronous speed in rev/min

ω_s is the synchronous speed in rad/s

n_r is the rotor speed in rev/min

ω_r is the rotor speed in rad/s

Another form of (7.3) often used is

$$\omega_r = \omega_s(1 - s) \tag{7.4}$$

The relative motion between the magnetic field and the rotor conductors induces a rotor voltage at a frequency called slip frequency, (f_r), which is related to the stator frequency, (f_s), by the expression

$$f_r = s \, f_s \qquad (7.5)$$

An easy way to remember (7.5) is to consider that at a blocked rotor, the rotor frequency is the supply frequency, where $f_r = (1.0)f_s$. Then when the rotor is moving at synchronous speed, the rotor frequency will be zero, where $f_r = 0.0 \times f_s = 0.0$.

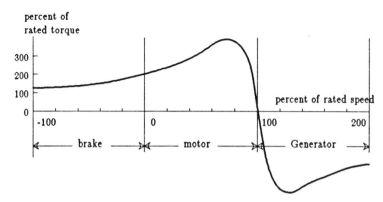

Figure 7.3 Typical torque-speed characteristic of a polyphase induction motor.

The torque-speed curve for a common induction machine is shown in Figure 7.3 for speeds between -100 percent and +200 percent of synchronous (+200 percent and -100 percent of slip). Theoretically, this curve could be extended to speeds of +infinity and -infinity with the practical limit established by the mechanical strength of the machine. A plot of this curve could be obtained by connecting a variable torque such as the waterwheel to the shaft of the induction machine. If water is allowed to flow over the waterwheel so as to oppose the machine torque, as shown in Figure 7.4a, the motor characteristics may be observed. There is an amount of water flow, that will provide a torque equal to the maximum torque available from the motor. If the waterwheel torque is increased beyond this maximum value, the motor will stall. Water flow should be adjusted to the stall torque. If the water flow is further increased, the induction machine will rotate in a reversed (or negative) direction in a braking mode, and the reverse speed will be limited only by the mechanical strength of the machine parts.

Now consider the case where the water flow assists the motor action of the induction machine, as shown in Figure 7.4b. For a small water flow, the waterwheel torque will increase so as to increase the induction machine speed toward synchronous speed and thus to compensate for the rotational losses. As the water flow is increased and the induction machine rotates faster than synchronous speed, the machine becomes a generator.

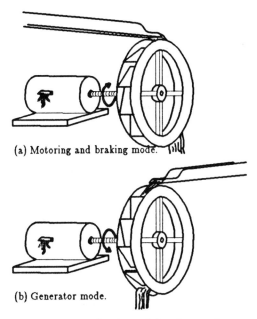

(a) Motoring and braking mode.

(b) Generator mode.

Figure 7.4 A system for determining the complete torque-speed curve for an induction machine.

When the machine is operating as a generator in parallel with a large system, the terminal voltage is determined by the system voltage and the power out is determined by the speed (or torque) of the prime mover. Also, the power factor of the energy is established by the speed and characteristics of the induction machine. However, when the induction machine is connected to an isolated load, the voltage is determined by the power factor of the load. The disadvantage of the induction generator is that it operates at a leading power factor at a speed greater than synchronous when used with a large interconnected power system. When operated as an isolated generator, the frequency is determined by the speed of the prime mover and the voltage is determined by the load power factor, which is adjusted by adding or subtracting capacitors in parallel with the load.

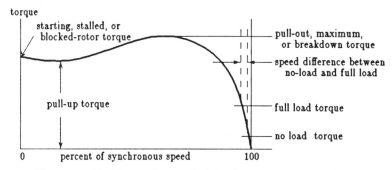

Figure 7.5 Significant points on the induction motor torque-speed curve.

The torque-speed curve showing the motor characteristics for the induction machine is shown in Figure 7.5. The important points on this curve include (1) starting torque (also called stall torque or blocked-rotor torque), (2) pull-up torque (the maximum load that can be started), (3) pull-out torque (also called maximum or breakdown torque), (4) full-load torque (the rated torque established by the manufacturer), (5) no-load torque (the torque represented by the no load friction, windage, and core losses).

7.2 EQUIVALENT CIRCUIT

A commonly used equivalent circuit that was developed from magnetic field and electric circuit equations (A29, p. 285) is shown in two forms in Figure 7.6 and is credited to Steinmetz. This circuit is the single-phase equivalent of the induction machine operated under balanced conditions and is similar to the power-tee equivalent circuit for the single-phase transformer. The differences are that a variable resistance representing the mechanical loading has been added, and the shunting resistance representing the core loss has been deleted. The core loss is included as part of the mechanical loading for ease in computation. The resistance $r_{2e}(1-s)/s$ in Figure 7.6b represents the gross mechanical load, including friction, windage, and the load connected to the motor shaft. Note that the equivalent mechanical load resistance plus the rotor resistance is simply r_{2e}/s, the resistance shown in Figure 7.6a. This means

$$\frac{r_{2e}(1-s)}{s} + r_{2e} = \frac{r_{2e}}{s} - r_{2e} + r_{2e} = \frac{r_{2e}}{s} \tag{7.6}$$

The quantitative solutions are of two types: Those related to determining the input quantities such as input current and power, and those concerned with the output quantities such as torque and shaft power.

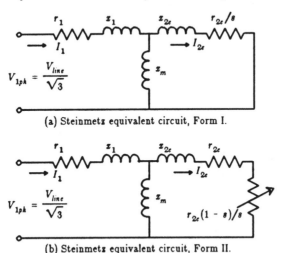

(a) Steinmetz equivalent circuit, Form I.

(b) Steinmetz equivalent circuit, Form II.

Figure 7.6 Steinmetz equivalent circuit for an induction motor.

A. Induction Motor Equations

There are several methods to evaluate machine performance used by the numerous authors and machine designers. It has been reported that even though the Steinmetz model of Figure 7.6 is an approximate circuit, the results are more realistic than many of the more refined equivalent circuits. Even though the procedure may appear different, the results are more accurate and the procedure is more simple. McPherson [A29] has shown a less complicated computational procedure for determining induction motor performance by a power flow concept. Begin with the equation for power input to the motor.

1. Power in

$$P_{in} = \sqrt{3}\,V_1 I_1 \cos\theta_1 = 3\,V_{1ph} I_1 \cos\theta_1 \qquad (7.7)$$

2. There are five categories of losses:

a. Stator winding loss for a three-phase machine is

$$\text{stator winding loss} = 3 I_1^2 r_1 \qquad (7.8)$$

b. The core loss is listed as P_{core} and for convenience in computation is included as part of the rotational losses.

c. Rotor winding loss for a three-phase machine is

$$\text{rotor winding loss} = 3I_2^2 r_2 \qquad (7.9)$$

d. Friction and windage loss $= P_{f\&w}$ is the sum of the bearing friction and the windage (or fan) reaction of the rotor and rotor cooling fins.

e. Stray load loss $= P_{stray}$ is caused by the magnetic fields in the end zone of the windings and harmonic magnetic fields. The stray loss is assumed to be 1.0 percent of motor output when not available by other tests.

f. The friction, windage, and stray losses included with the hysteresis and eddy current losses are called the rotational losses.

$$P_{rot} = \text{rotational losses} = P_{f\&w} + P_{core} + P_{stray} \qquad (7.10)$$

3. Gap power. A convenient quantity in determining motor performance is the power transferred across the air gap and is the input power less the stator winding loss, thus

$$P_g = (\text{power crossing the air gap}) = P_{in} - \text{stator winding loss}$$

$$P_g = \sqrt{3}\,V_1 I_1 \cos\theta_1 - 3I_1^2 r_1 = 3I_1^2 R_f \qquad (7.11)$$

4. Developed mechanical power. The air gap power is partly consumed as rotor winding loss, and the remainder is the gross mechanical power (also called developed mechanical power). Thus,

$$P_{gross} = P_g - 3I_2^2 r_2 = P_g(1 - s) \qquad (7.12)$$

5. Finally, the useful power available at the shaft is the gross mechanical power less the rotational losses.

$$P_{shaft} = P_{gross} - (P_{f\&w} + P_{core} + P_{stray}) \qquad (7.13)$$

$$= P_{gross} - P_{rot}$$

6. The torque available to the machine shaft is

$$\tau_{shaft} = \frac{P_{shaft}}{\omega_r} = \frac{P_{gross} - P_{rot}}{\omega_s(1 - s)} \quad \text{N·m} \qquad (7.14)$$

There is a torque called the *developed torque* (also called internal torque or electromagnetic torque) that is developed by the electromagnetic energy conversion process and is defined by

$$\tau_d = \frac{P_{gross}}{\omega_r} \tag{7.15}$$

Note that $\omega_r = \omega_s(1 - s)$ by (7.4); thus, (7.15) can be written as

$$\tau_d = \frac{P_{gross}}{\omega_s(1 - s)} = \frac{P_g(1 - s)}{\omega_s(1 - s)} = \frac{P_g}{\omega_s} \tag{7.16}$$

This means that three of the motor values may be determined directly from the air gap power.

$$\text{rotor winding loss} = sP_g \tag{7.17}$$

$$P_{gross} = (1 - s)P_g \tag{7.18}$$

$$\tau_d = \frac{P_g}{\omega_s} \tag{7.16}$$

B. Induction Motor Performance Calculations

Use the equations of Section 7.2A to calculate the motor performance. The following steps offer a systematic procedure.

1. Calculate the synchronous speed $n = \dfrac{120\, f}{P}$ where P = number of poles.

2. Calculate or establish a value for the slip

$$s = \frac{n_s - n_r}{n_s}$$

3. Input impedance

$$\mathbf{Z}_{in} = (r_1 + jx_1) + \mathbf{Z}_f$$

where

$$\mathbf{Z}_f = (r_{2e}/s + jx_{2e}) \text{ in parallel with } jx_m$$

$$\mathbf{Z}_f = \frac{(r_{2e}/s + jx_{2e})jx_m}{r_{2e}/s + j(x_{2e} + x_m)}$$

4. The stator current is then:

$$\mathbf{I}_1 = \frac{\mathbf{V}_{1ph}}{\mathbf{Z}_{in}}$$

This means that the input power is:

$$P_{in} = 3V_{1ph}I_1\cos\theta_1 = \sqrt{3}V_1I_1\cos\theta_1$$

where θ_1 is the angle related to the input impedance, $\mathbf{Z}_{in}$.

5. Stator winding loss $= 3I_1^2 r_1$

6. Air gap power $= P_g = 3I_2^2 r_2/s = 3I_1^2 R_f$

7. Rotor winding loss $= 3I_2^2 r_2 = sP_g$

8. Gross developed mechanical power

$$P_{gross} = P_g - 3I_2^2 r_2 = 3I_2^2 r_2(1 - s)/s = P_g(1 - s)$$

9. Developed torque $= T_d = P_g/\omega_s$

10. Net shaft power $= P_{shaft} = P_{gross} - P_{rot}$

$$P_{shaft} = P_{gross} - (P_{f\&w} + P_{core} + p_{stray}) \quad \text{and} \quad HP = \frac{P_{shaft}}{746}$$

11. Net shaft torque $= \tau_{shaft} = P_{shaft}/\omega_r$

12. Motor efficiency $= E\!f\!f = P_{shaft}/P_{in}$

Example 7.1

Motor performance characteristics for a slip of 4 percent. Given a three-phase induction motor, 350 hp, 1100 rev/min, 2300 V, design B, with

$r_1 = 0.15 \ \Omega, \ r_{2e} = 0.95 \ \Omega,$

$x_1 = x_{2e} = 1.63 \ \Omega, \ x_m = 81.4 \ \Omega,$

$P_{f\&w} + P_{core} = 8300 \ W$

Assume $P_{stray} = 1$ percent of P_{gross}

Determine the following for a rotor slip of 4 percent: (a) the stator current, input power, and input kVA; (b) the shaft power output and torque; (c) the efficiency.

Solution

1. Calculate the synchronous speed. Note that the motor is rated 1100 rev/min, which implies the motor is 6 poles with a synchronous speed of 1200 rev/min. This means

$$\omega_s = 1200 \ (rev/min)(2\pi \ rad/rev)(min/60s)$$

$$= 1200(2\pi/60) = 125.7 \ rad/s$$

2. The slip is given as 4 percent (or 0.04 per unit), which is a speed of $n_r = n_s(1 - s) = 1200(1 - 0.04) = 1152 \ rev/min$.

3. The input impedance is obtained by combining the field impedance and the stator impedance.

$$\mathbf{Z}_f = \frac{(r_{2e}/s + jx_{2e})jx_m}{r_{2e}/s + j(x_{2e} + x_m)} = \frac{(0.95/0.04 + j1.63)j81.4}{0.95/0.04 + j(1.63 + 81.4)}$$

$$= (21.10 + j7.63) \ \Omega$$

$$\mathbf{Z}_{in} = (r_1 + jx_1) + \mathbf{Z}_f = (0.15 + j1.63) + (21.10 + j7.63)$$

$$= 21.25 + j9.26 = 23.18\underline{/23.55^\circ} \ \Omega$$

4. The input current is

$$I_1 = \frac{V_{1ph}}{Z_{in}} = \frac{2300/\sqrt{3}}{23.18} = 57.28 \ A$$

with an input power of

$$P_{in} = \sqrt{3}V_1 I_1 \cos\theta_1 = \sqrt{3}(2300)(57.28) \cos 23.55^\circ = 209.2 \ kW$$

5. The stator winding loss is

stator winding loss $= 3I_1^2 r_1 = 3(57.28)^2(0.15) = 1.48 \ kW$

6. The air gap power is

$$P_g = 3I_1^2 R_f = 3(57.28)^2(21.10) = 207.7 \ kW$$

7. The rotor winding loss is

rotor winding loss $= sP_g = (0.04)(207.7) = 8.309 \ kW$

8. The developed mechanical power is

$$P_{gross} = P_g(1 - s) = 207.7(1 - 0.04) = 199.4 \ kW$$

9. The developed torque is

$$\tau_d = \frac{P_g}{\omega_s} = \frac{207{,}700}{125.7} = 1{,}653 \text{ N·m}$$

10. The net mechanical shaft power is

$$P_{shaft} = P_{gross} - P_{rot} = P_{gross} - \left(P_{f\&w} + P_{core} + P_{stray}\right)$$

$$= 199.4 - (8.300 + 0.01(199.4)) = 199.4 - 10.3$$

$$= 189.1 \text{ kW} \quad \text{or} \quad \frac{P_{shaft}}{0.746} = 253.5 \text{ hp}$$

11. $\tau_{shaft} = \dfrac{P_{shaft}}{\omega_r} = \dfrac{189{,}100}{120.6} = 1568 \text{ N·m}$

where $\omega_r = 1152(2\pi/60) = 120.6 \text{ rad/s}$

12. $Eff = \dfrac{P_{shaft}}{P_{in}} = \dfrac{189.1}{209.2} = 0.904 \text{ or } 90.4\%$

Example 7.2

Motor performance characteristics for starting conditions. Given the motor of Example 7.1, determine the following for the starting condition: (a) stator current, (b) input power, (c) starting developed torque, (τ_d).

Solution

1. The synchronous speed from Example 7.1 is 1200 rev/min or 125.7 rad/s.

2. The slip at starting is 1.0 per unit.

3. The input impedance is obtained by combining the field impedance and the stator impedance where the slip is 1.0 per unit.

$$\mathbf{Z}_f = \frac{(r_{2e}/s + jx_{2e})jx_m}{r_{2e}/s + j(x_{2e} + x_m)} = \frac{(0.95/1.0 + j1.63)j81.4}{0.95/1.0 + j(1.63 + 81.4)}$$

$$= 0.913 + j1.608 = 1.849\underline{/60.42^\circ}$$

$$\mathbf{Z}_{in} = (r_1 + jx_1) + \mathbf{Z}_f = (0.15 + j1.63) + (0.913 + j1.608)$$

$$= 1.063 + j3.238 = 3.408\underline{/71.83^\circ}$$

4. The input current is

$$I_1 = \frac{V_{1ph}}{Z_{in}} = \frac{2300/\sqrt{3}}{3.408} = 389.6 \text{ A}$$

with an input in kVA of

$$S = \sqrt{3}(2300)(389.6) = 1552 \text{ kVA}$$

and with an input power of

$$P_{in} = S_{in} \cos \theta_1 = 1552 \cos 71.83° = 484 \text{ kW}$$

5. The air gap power is

$$P_g = 3I_1^2 R_f = 3(389.6)^2(0.913) = 416 \text{ kW}$$

6. The developed starting torque is

$$\tau_d = \frac{P_g}{\omega_s} = \frac{416,000}{125.7} = 3310 \text{ N·m}$$

C. Maximum Torque and Power

Two important quantities that explain induction motor performance are (1) the maximum torque and (2) the slip at maximum torque. The equations used in the previous sections do not directly lead to the desired form of torque equation. For this reason, consider the cantilever approximate circuit of Figure 7.7, which is a circuit similar to the transformer cantilever circuit. Begin by solving for the rotor current.

$$I_2 = \frac{V_1}{(r_1 + r_2/s) + j(x_1 + x_2)}$$

$$I_2 = \frac{V_1}{\sqrt{(r_1 + r_2/s)^2 + (x_1 + x_2)^2}} \tag{7.19}$$

Next, find the gross developed mechanical power:

$$P_{gross} = 3I_2^2 r_2(1 - s)/s = \frac{3V_1^2 r_2(1 - s)/s}{(r_1 + r_2/s)^2 + (x_1 + x_2)^2} \tag{7.20}$$

The developed torque is then:

$$\tau_d = \frac{P_{gross}}{\omega_r} = \frac{P_{gross}}{\omega_s(1-s)} = \frac{3V_1^2 r_2/s}{\omega_s(r_1 + r_2/s)^2 + (x_1 + x_2)^2} \qquad (7.21)$$

The maximum torque is obtained by taking the derivative of the torque with respect to the slip and then setting the derivative to zero. Thus,

$$\frac{d\tau_d}{ds} = \left(\frac{3V_1^2 r_2}{\omega_s}\right)\left[\frac{((r_1 + r_2)^2 + (x_1 + x_2)^2)(-s^{-2}) - (2/s)(r_1 + r_2/s)(-r_2 s^{-2})}{(r_1 + r_2/s)^2 + (x_1 + x_2)^2}\right]$$

Then for $\dfrac{d\tau_d}{ds} = 0$,

$$0 = \frac{-(r_1 + r_2/s)^2 + (x_1 + x_2)^2}{s^2} + (2r_2/s^3)(r_1 + r_2/s)$$

$$0 = -r_1^2 + \left(\frac{r_2}{s}\right)^2 - (x_1 + x_2)^2$$

$$s_{max\ T} = \frac{r_2}{\sqrt{r_1^2 + (x_1 + x_2)^2}} \qquad (7.22)$$

The maximum torque is found by combining (7.22) with (7.21) for

$$\tau_{max} = \frac{3V_{1ph}^2}{2\omega_s\left[r_1 + \sqrt{r_1^2 + (x_1 + x_2)^2}\,\right]} \qquad (7.23)$$

Note that the magnitude of maximum torque is independent of the rotor resistance, while the slip at maximum torque is directly related to the rotor resistance.

Maximum power output

A second quantity often referred to is that of maximum gross power output. This is readily found by using the maximum power transfer theorem where

$$r_2(1-s)/s = \sqrt{(r_1 + r_2)^2 + (x_1 + x_2)^2} = Z \qquad (7.24)$$

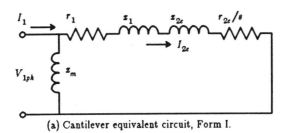

(a) Cantilever equivalent circuit, Form I.

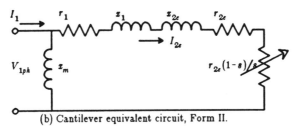

(b) Cantilever equivalent circuit, Form II.

Figure 7.7 Cantilever equivalent circuit.

Rearranging terms produces:

$$r_2 - sr_2 = sZ$$

or

$$\frac{r_2}{r_2 + Z} = s$$

$$S_{max\ P} = \frac{r_2}{r_2 + \sqrt{(r_1 + r_2)^2 + (x_1 + x_2)^2}} \tag{7.25}$$

Example 7.3

For Example 7.1, determine the slip when the machine is developing maximum torque and also the magnitude of the maximum torque.

$$s_{max\ T} = \frac{0.95}{\sqrt{(0.15)^2 + (1.63 + 1.63)^2}} = 0.291 \quad \text{per unit slip}$$

$$\tau_{max} = \frac{3(1328)^2}{(2(125.7)(0.15) + \sqrt{(0.15)^2 + (1.63 + 1.63)^2}} = 6165 \ \text{N·m}$$

7.3 CIRCLE DIAGRAM

Most early writers of electric machine texts described a graphical solution of induction machine performance called the circle diagram. Since the results are not accurate for all values of slip — especially for the double squirrel-cage type of rotor — this procedure will not be explained in detail. However, there are some important observations that are not obvious from other methods.

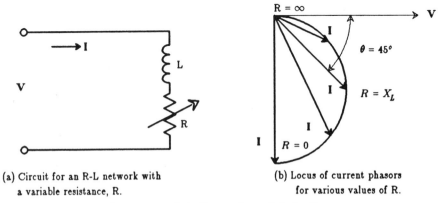

(a) Circuit for an R-L network with a variable resistance, R.

(b) Locus of current phasors for various values of R.

Figure 7.8 Circle diagram for an R-L circuit.

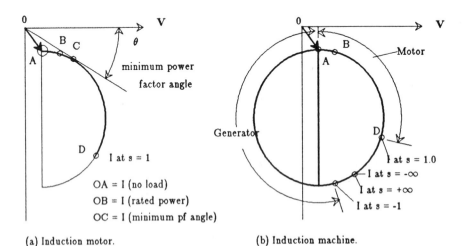

(a) Induction motor.

OA = I (no load)
OB = I (rated power)
OC = I (minimum pf angle)

(b) Induction machine.

Figure 7.9 Simplified circle diagram for an induction machine.

The circle diagram is a plot of the induction motor stator current for all values of load. Begin by considering a plot of current in a simple R-L series circuit where R is varied between zero and infinity, as shown in Figure 7.8. A careful examination of the equivalent circuit for an induction motor, (Figure 7.7), reveals that the induction motor is a parallel R-L circuit. The circle diagram for the induction motor is shown as Figure 7.9a. A brief review of the diagram reveals that (a) the power factor is always lagging (note the minimum power factor angle), (b) the starting current that occurs when the slip is equal to 1.0 per unit is large compared to rated current and is at a large lagging power factor angle, and (c) the no load power factor is small (large power factor angle). The locus of current phasors may be plotted for all values of slip between plus and minus infinity. The plot of torque vs. speed in Figure 7.3 agrees with the ideas obtained from the circle diagram of Figure 7.9b.

7.4 WOUND-ROTOR INDUCTION MOTOR

From (7.23), the maximum torque of an induction motor is independent of the rotor resistance, r_{2e}. However, it will also be noted that the slip at maximum torque is directly proportional to the rotor resistance, as indicated in (7.22). Thus, if the rotor conductors are insulated and connected to slip rings so that the rotor can be connected to an external variable resistance (see Figure 11.18), the torque-speed characteristics may be varied as noted in Figure 7.10 to allow for (a) maximum starting torque, (b) maximum running efficiency, and (c) variable speed for a given load torque. The initial cost and maintenance of the wound-rotor motor is greater than for the squirrel-cage motor.

The I^2R loss in the external resistance for the wound-rotor machine is undesirable. The development of electronic and mechanical methods of squirrel-cage motor speed control have caused a loss of popularity for the wound-rotor machine. During the first half of the twentieth century, several interesting methods were developed that reduced the rotor winding losses. These methods used other machines that generated opposing voltages to produce the effect of external resistance while not increasing the losses.

The wound-rotor machine may also be used as (a) a three-phase synchro (see Figure 12.8) or (b) as a variable frequency generator. The latter feature is accomplished by driving the rotor with a variable speed machine. Then the rotor will supply an energy source with a frequency that is the difference between the stator frequency and the rotor speed.

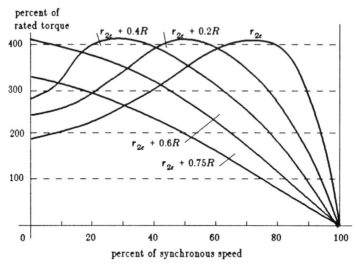

Figure 7.10 Wound-rotor induction motor torque-slip curves showing the effect of changing the rotor circuit resistance.

7.5 SQUIRREL-CAGE ROTOR DESIGN

The rotors of a squirrel-cage induction motor can be designed so that the torque-speed curve fits any of the several characteristics called "Designs," as shown in Figure 7.11. Note that the squirrel-cage rotor is formed with laminations of ferromagnetic material bolted together, as shown in Figure 7.12. The blank spaces are filled with molten aluminum or with rods of conducting material such as copper and aluminum to form the completed rotor assembly such as that shown in Figure 7.1.

The Design A induction motor is designed to have a minimum change in speed between no load and full load. This is accomplished by constructing the rotor cage of low resistance conductors as required by (7.22) and shown in Figure 7.12a. The disadvantage of this design is that the starting currents are large and usually require reduced voltage starting.

The Design F squirrel-cage rotor of Figure 7.12f takes into account the change of rotor leakage reactance with respect to the speed of the rotor. At low values of slip, the frequency of the rotor current is near 1 to 2 Hz, per (7.5), and since the rotor bars are relatively large the rotor resistance is low and the motor operates with good efficiency and low slip for rated load. The frequency of the rotor current at starting is the supply frequency of 60 Hz, as in (7.5). Because the Design F rotor bars are embedded

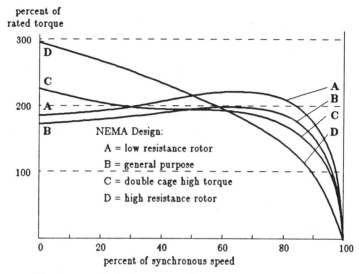

Figure 7.11 Typical torque-speed curves for a squirrel-cage induction motor showing Design A, B, C, and D type rotors.

in the rotor, the leakage inductance is high. Also, the leakage inductance is nonuniform, as shown in Figure 7.13. The nonuniform flux produces the effect of a higher rotor resistance at the starting current frequency. The final result is a motor that has much lower starting current than Design A with an acceptable starting torque and good efficiency and low slip at full load, which is well-suited to large fan motors. This is not an officially accepted design by NEMA, and it is mentioned here because it points out the embedded rotor bar phenomenon.

The Design D induction motor is suitable for high starting torque applications. This is accomplished by constructing the rotor cage of high resistance conductors as indicated by (7.5) and shown in Figure 7.12d. The running torque of this design is relatively low. However, this is acceptable for applications such as traction motors for use with electric carts, electric automobiles, etc.

The general purpose Design B squirrel-cage induction motor provides good efficiency with low slip and good starting torque without excessive starting currents. This is accomplished with the deep bar configuration of Figure 7.12b and Figure 7.13b. This motor design is used for more applications than all other designs put together.

Finally, Design C uses the double squirrel cage of Figures 7.12c and 7.13c. This produces high starting torque with the outer cage at 60 Hz

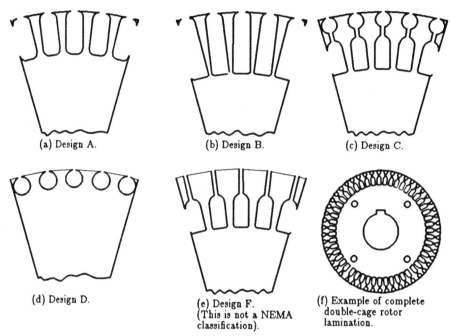

(a) Design A.   (b) Design B.   (c) Design C.

(d) Design D.   (e) Design F.
(This is not a NEMA
classification).   (f) Example of complete
double-cage rotor
lamination.

Figure 7.12 Rotor lamination slot designs for squirrel-cage induction motors. Figures *a* through *e* are sections of laminations only.

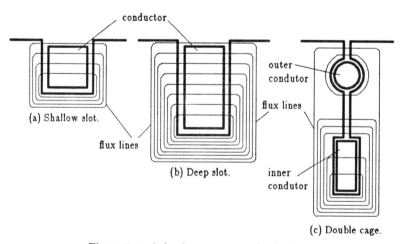

conductor

(a) Shallow slot.

flux lines

flux lines

(b) Deep slot.

outer
condutor

flux lines

inner
condutor

(c) Double cage.

Figure 7.13 Induction motor rotor flux leakage.

rotor frequency and the high running torque of the combined cages at the running rotor frequencies of 1 to 2 Hz. This design is used in conveyer systems and other systems requiring relatively high starting and running torques.

In summary, the squirrel-cage induction motor may have any of several speed-torque characteristics by preselecting a specific rotor design as shown in Figures 7.11 and 7.12.

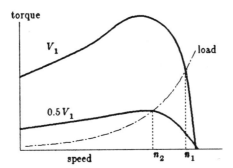

Figure 7.14 Speed control by means of line voltage.

7.6 SPEED CONTROL

The induction motor fulfills admirably the requirements of a substantially constant speed drive. Many motor applications, however, require multiple speed or adjustable speed ranges. Below are listed some of the methods used to obtain variable speed.

1. Change number of poles.

2. Vary terminal voltage (Figure 7.14).
    a. Has limited value for Design A, B, and C induction motors.
    b. Excellent with SCR control for Design D motors.

3. Use mechanical methods.
    a. Gears.
    b. Pulleys.
    c. Variable-pitch pulley system; (brand names include US Motors' Varidrive, GE's Polydyne, Louis-Allis' Allispeed (Figure 7.15).
    d. Hydraulic pump and motor (Figure 7.16).

4. Vary the frequency (Figure 7.17).

5. Use electromechanical methods.
    a. Ward-Leonard system (Figure 9.25).
    b. Hysteresis clutch.
    c. Eddy-current clutch; brand names include Louis-Allis' Adjusto
       Speed and GE's Kinatrol.

6. Speed control for wound-rotor machines [see C4, p320].
    a. Vary the external rotor resistance (11.18).
    b. Le Blanc.
    c. Kramer.
    d. Modified Kramer.
    e. Scherbius.
    f. Schrage brush shift.
    g. Concatenation (Tandem or cascade control).

The following is a description of several speed control systems.

A. Variable-pitch Pulley

The mechanical transmission system consists of a V-belt and adjustable-pitch pulleys (Figure 7.16). Adjustable-pitch drive output speed is controlled mechanically by a belt transmission system consisting of two reciprocally adjustable pulleys, (A,B), mounted on a splined driving shaft, (C), and on a splined driven output shaft, (D), and connected by a V-belt of constant circumference, (E).

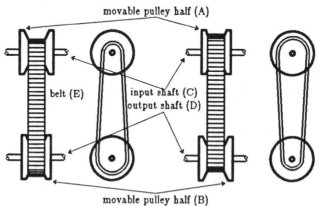

(a) Low output speed.   (b) High output speed.

Figure 7.15 Adjustable speed of an induction motor by a variable-pitch pulley system.

Each pulley consists of two halves. One half of each is permanently fixed to its respective shaft while the other half is free to move axially on its shaft. Position of the movable pulley half on the driving shaft is regulated by a control handwheel that operates a crank-and-screw mechanism, which in turn directs an adjustment arm against the movable pulley half. All pulley faces are sloping, and movement of the driving pulley half causes the effective pulley diameter to vary as the belt rides closer to or further from the drive shaft. The pitch diameter on the output pulley varies reciprocally. The free half of the output pulley moves against a compensating spring. The spring automatically adjusts belt tension to provide the transmission of rated torque.

Output speed is determined by the diameter of the driven pulley. As the diameter of the driving shaft pulley increases, the belt "rides" toward the pulley's outer perimeter. This action pulls the belt closer to the driven pulley shaft, reducing the effective diameter and causing an increase in output speed. Conversely, as adjustment arm pressure is removed, the compensating spring on the driven shaft closes the pulley halves. This action increases the effective pulley diameter and causes a reduction in output speed.

B. Hydraulic Pump and Motor

A commonly used hydraulic servomechanism power element employs a variable displacement pump in which the quantity of oil pumped is proportional to the displacement of the pump control from the neutral, and the direction of oil flow is determined by the direction of the displacement of the pump tilt plate from neutral. Figure 7.16 indicates such a hydraulic pump and motor combination. The oil from the pump is supplied by a hydraulic motor that has a speed proportional to the magnitude and direction of the oil flow to it.

C. Variable Frequency

Figure 7.17 shows a speed changing system that operates on the principle of varying frequency. This is accomplished by driving a synchronous generator with a variable pitch belt system (see section 7.6A). It should be noted that the supply voltage must be reduced as the frequency is reduced.

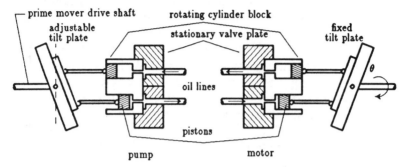

Figure 7.16 Obtaining variable shaft speed by driving a hydraulic pump and motor system with an induction motor.

Figure 7.17 Variable frequency speed control system. *(Courtesy of U.S. Electrical Motors Division of Emerson Electric Company)*

7.7 INDUCTION MOTOR DIFFICULTIES

Single-phasing is the opening of one of the supply lines to a three-phase induction motor. It will be shown in Chapter 10 that the machine delivers mechanical torque in this single-phase mode. However, the current in the remaining two motor stator coils is increased above the standard three-phase current for a given shaft load. This overcurrent is often not sufficient to trip the "overload protection devices" but enough to cause damage to the motor.

7.8. PERFORMANCE TESTS (ADVANCED TOPIC)

The impedance values for the induction motor equivalent circuit can be found by a set of tests similar to those for the two-winding transformers. These tests are referred to as the *no-load* (or running-light) test and the *blocked rotor* test. A difficulty that arises, however, is that the frequency of the rotor current is not constant for all operating modes. The rotor current frequency for starting is the supply frequency, while the rotor current frequency at rated load is but a fraction of the supply frequency. This means that the motor impedances are not the same for all operating values. The industry has established the practice of using two sets of blocked rotor data. One blocked rotor test conducted at rated frequency provides data for calculations relating to starting conditions, and the other blocked rotor test conducted at 15 Hz provides data for calculations relating to parameters for rated speed operation. The IEEE Standard 112-1978 [O.5(j)] lists the procedure for these tests. However, since the method is rather cumbersome and employs the iterative process, the following method shown by Fitzgerald and Kingsley (A14) provides a reasonable alternative for the novice.

A. Blocked Rotor Test at 60 Hz[1]

The rotor is clamped in the locked position, hence the names blocked rotor, locked rotor, stalled rotor, etc. An adjustable voltage source is connected to the motor and adjusted for the approximately rated stator current. Measurements are made of the voltage, current, and power. The following equations are used to determine motor parameters. The 60 Hz data is for use in calculating starting torques:

$$R_L = \frac{(P_L/3)}{I_L^2} \tag{7.26}$$

$$Z_L = \frac{(V_L/\sqrt{3})}{I_L} \tag{7.27}$$

---

[1]Subscripts:

    L = quantities pertaining to blocked rotor tests

    o = quantities pertaining to no-load test

    M = quantities pertaining to the magnetizing circuit

    1 = quantities pertaining to the stator circuit

    2 = quantities pertaining to the rotor circuit

$$X_L = \sqrt{Z_L^2 - R_L^2} \qquad (7.28)$$

B. Blocked Rotor Test at 15 Hz

The measurements of this test are made with a supply frequency of 15 Hz and are used in low slip calculations.

$$R_{L15} = \frac{P_{L15}/3}{I_{L15}^2} = R_{L60} \qquad (7.29)$$

$$Z_{L15} = \frac{V_L/\sqrt{3}}{I_{L15}} \qquad (7.30)$$

$$X_{L15} = \sqrt{Z_{L15}^2 - R_{L15}^2} \qquad (7.31)$$

$$X_{L60} = X_{L15}\left(\frac{60 \text{ Hz}}{15 \text{ Hz}}\right) \qquad (7.32)$$

$$\mathbf{Z}_{L60} = R_{L60} + jX_{L60} \qquad (7.33)$$

C. No Load Test

This test is performed with rated terminal voltage and with the rotor free to rotate without any mechanical load. Again, measure the voltage, current, and power.

$$R_o = \frac{P_o/3}{I_o^2} \qquad (7.34)$$

$$Z_o = \frac{V_o/\sqrt{3}}{I_o} \qquad (7.35)$$

$$X_o = \sqrt{Z_o^2 - R_o^2} \qquad (7.36)$$

D. Resistance Test

$$r_1 = \text{dc resistance per phase} \qquad (7.37)$$

E. Calculations

1. For stalled rotor constants:

$$X_L = X_1 + X_2 \qquad (7.38)$$

with the proportions for X_1 and X_2 as shown in Table 7.1.

Table 7.1
Squirrel-cage Induction Motor
Impedance Ratios

Design of Motor	Fraction of $z_l + z_2 = z_L$	
	z_1	z_2
Design A	0.5	0.5
B	0.4	0.6
C	0.3	0.7
D	0.5	0.5
Wound Rotor	0.5	0.5

$$X_M = X_o - X_1 \qquad (7.39)$$

$$r_2 = (R_L - r_1) \left[\frac{X_M + X_2}{X_M} \right]^2 \qquad (7.40)$$

2. For speeds near rated load:

$$X_{L60} = X_{1(60)} + X_{2(60)} \qquad (7.41)$$

with the proportion for $X_{1(60)}$ and $X_{2(60)}$ as given in the Table 7.1.

$$X_{M60} = X_o - X_{1(60)} \qquad (7.42)$$

$$r_{2(60)} = (R_{L60} - r_1) \left[\frac{X_{M60} + X_{2(60)}}{X_{M\,60}} \right]^2 \qquad (7.43)$$

F. Proof for Equations 7.40 and 7.43

For the given input power at locked rotor

$$P_{in} = 3I_L^2 R_L \qquad (7.44)$$

the power dissipated in the stator is

$$P_{stator} = 3I_L^2 r_1 \tag{7.45}$$

and the power dissipated in the rotor is

$$P_{rotor} = 3I_{2L}^2 r_2 \tag{7.46}$$

The locked rotor current by current division is

$$\mathbf{I}_{2L} = \frac{\mathbf{I}_L (jX_M)}{r_2 + j(X_M + X_2)}$$

$$\approx \mathbf{I}_L \left[\frac{X_M}{X_M + X_2} \right] = \mathbf{I}_L \left[\frac{X_M}{X_{22}} \right] \tag{7.47}$$

where

$$X_{22} = X_M + X_2$$

Equation 7.47 may be rewritten as

$$\frac{I_{2L}}{I_L} = \frac{X_M}{X_{22}} \tag{7.48}$$

Combine (7.42), (7.43), and (7.44) so that

$$P_{in} = P_{stator} + P_{rotor}$$

or

$$I_L^2 R_L = I_l^2 r_1 + I_{2L}^2 r_2 \tag{7.49}$$

Divide by I_L^2; then

$$R_L = r_1 + \left(\frac{I_{2L}}{I_L} \right)^2 r_2 \tag{7.50}$$

But the current ratio is defined by (7.48); thus

$$R_L = r_1 + \left(\frac{X_M}{X_{22}} \right)^2 r_2 \tag{7.51}$$

Finally, solve for r_2

$$r_2 = (R_L - r_1)\left(\frac{X_M + X_2}{X_M}\right)^2 \tag{7.52}$$

PROBLEMS

7.1 Name the major parts of a squirrel-cage induction motor and describe their functions.

7.2 How do you change the direction of rotation of a three-phase induction motor?

7.3 What is meant by the expressions: (a) pull-out or breakdown torque, (b) starting torque, (c) stalled torque, (d) rated torque?

7.4 List the synchronous speeds for 60 Hz squirrel-cage induction motors having 2, 4, 6, 8, 10, 12, and 14 poles.

7.5 What is the maximum theoretical speed of a 200 V, three-phase, 60 Hz, 15 hp, Design B, induction motor supplied from a 60 Hz, 208 V source?

7.6 A 50 hp, 460 V, 60 Hz, Design B, three-phase induction motor requires a full load current of 65 A at an 85 percent power factor. What is the motor efficiency at full load?

7.7 What is the kVA input to a 30 hp, three-phase induction motor operating at rated output with 88 percent power factor and 85 percent efficiency?

7.8 Is a squirrel-cage motor generally considered a constant speed or variable speed motor? Explain.

7.9 What is a double squirrel-cage rotor? What are the advantages and disadvantages of such a rotor?

7.10 Given a 100 hp, three-phase, Design B, 60 Hz, 873 rev/min induction motor, (a) how many poles exist for this motor? (b) what is the percent slip at full load? (c) what is the frequency of the rotor voltages? (d) what is the speed in rev/min of (1) the rotor field with respect to the stator field? (2) the rotor field with respect to the rotor? and (3) the rotor field with respect to the stator? (e) what speed would the rotor have for a slip of 6 percent? (f) what is the rotor frequency at this speed? (g) Repeat parts (c) and (d) for a slip of 6 percent.

Table 7.2 Three-phase Squirrel-cage Induction Motor Data

	HP	Voltage	Design	Hz	FL Speed	SF†	Code‡	FLA	R_1	R_2	X_1	X_2	X_m	Losses rot§	Losses stray
A	3/4	200-230/460	B	60	1145	1.15	R	?/1.55							
B	2	200-230/460	B	60	3495	1.15	L	?/2.8							
C	2	200-230/460	B	60	1155	1.15	L	?/3.3							
D	20	200-230/460	B	60	1755	1.15	G	?/27.0							
E	20	200-230/460	C	60	1200*	1.00	G	?/26.6							
F	100	200-230/460	D	60	1800*	1.00		?/118							
G	100	200-230/460	C	60	1200*	1.00	G	?/116							
H	100	200-230/460	B	60	880	1.15	G	?/125							
J	200	460	B	60	1175	1.15		?/241							
K	400	460	B	60	1760	1.15									
L	2250	6600		60	3571	1.15		168							
M	1000	6600		60	890	1.15		88							
N	50	460	B	60	1775	1.15	A	61.1	.06	.054	.134	.201	13	1200	700
P	30	230	B	60	1710	1.0									
Q	20	200	B	60	1710	1.15	F	62.1	.118	.102	.166	.250	10.1	800	500
R	200	460	B	60	1725				.042	.038	.201	.302	9.06	5570	1400
S	200	2300	B	60	1725		E		1.4	.94	5.6	8.40	245	5960	1500

Notes

* -- synchronous speed
†SF -- Service Factor
‡Code -- See NEC Table 430-7(b) listed in Appendix B.
§rot = rotational loss = $P_{f\&w} + P_{core} + P_{stray}$

7.11 A 50 hp, 1725 rev/min, 60 Hz, three-phase induction motor is operating at rated speed. (a) What is the percent slip? (b) What is the frequency of the voltage induced in the rotor windings?

7.12 Determine the following for the induction motors of Table 7.2 operating at the percent slip shown below: (a) the stator current (I_1), (b) the power input, and (c) the input kVA.

Machine	Percent Slip
N	1.3
Q	5.0
R	4.0
S	4.17

7.13 Calculate the following for the induction motor of Problem 7.12: (a) the air gap power (P_g), (b) the gross mechanical power, (c) the output shaft power in kW, (d) the output shaft power in hp, and (e) the efficiency.

7.14 Calculate the following for the induction motor of Problem 7.12: (a) the developed torque and (b) the shaft torque.

7.15 Determine the following for the induction motors of Table 7.2 in the starting mode: (a) stator current (I_1), (b) input kVA, (c) input power, (d) air gap power (P_g), and (e) developed torque (τ_d).

7.16 Repeat Problems 7.12, 7.13, and 7.14 for the following values of slip:

Machine	Percent Slip
N	1.0
Q	4.0
R	3.0
S	3.0

7.17 What determines the maximum size of a squirrel-cage motor that can be started across the line?

7.18 A polyphase induction motor used to drive a ship has a variable frequency supply ranging from 40 to 90 Hz. A pole-changing switch is used to change the number of poles to either 12 or 18 poles. What are the minimum and maximum speeds of the motor?

7.19 What are the limitations of operating a 230 V three-phase induction motor from a 208 V source?

7.20 Describe the mechanical differences between a squirrel-cage induction motor and wound-rotor induction motor.

7.21 What are the major advantages and disadvantages of squirrel-cage induction motors?

7.22 What are the major advantages and disadvantages of wound-rotor induction motors?

7.23 What size three-phase transformer (in kVA) is required to supply a 1777 rev/min, 60 Hz, 460 V, three-phase, Design B, squirrel-cage induction motor? Sketch the circuit diagram. Assume full load slip of 1.3 percent. The motor parameters are:

$r_1 = 0.06$, $r_2 = 0.054$

$x_1 = 0.134$, $x_2 = 0.201$, $x_m = 13.0$

no load and stray losses = 1200 W

8

POLYPHASE
RECTIFICATION

8.1 INTRODUCTION

There are many applications in which direct current is needed. At one time direct current was supplied by a dc generator driven by an ac motor. The process of converting ac to dc can be accomplished for a fraction of the initial cost and with less maintenance by using rectifier circuits. A rectifier is a device that acts with electrical current flow in a manner similar to a check value in a water flow system.

The first rectifiers were used for relatively low power circuits and included the following types:

electrolytic

copper oxide

selenium

cold cathode

hot cathode — high vacuum

hot cathode — gas filled

The earlier large power rectifiers used a mercury arc. These included:

mercury tank rectifiers

ignitrons

thyratrons

valves

One of the larger rectifiers is the mercury arc valve used at the terminals of the 750 kV dc line between Oregon and California. Each valve is rated 125,000 V at 1250 A.

Beginning in the 1960s, the solid state devices proved to be less expensive, more efficient, and more reliable. These now include:

silicon diodes

thyristors
       silicon control rectifiers (SCR)
       silicon control switches (SCS)
       triacs

The electronic theory of the solid state P-N junction may be found in one of the many excellent texts on electronics and will not be included in this text. For this discussion the ideal diode will be used. The ideal diode offers zero impedance (short circuit) in the forward flow direction and infinite impedance (open circuit) in the reverse direction, as indicated in Figure 8.1a.

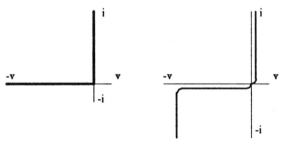

(a) Ideal diode characteristic.   (b) Silicon diode characteristic.

Figure 8.1 Diode characteristics.

There are numerous rectifier circuits of which the following are the more common. Those marked with an asterisk are described in some detail in this chapter. A collection of rectifier constants is given as Figure 8.2.

A. Rectifier Circuits

Single-phase
*   half-wave (Figure 8.3)

		Single Phase			Three-Phase		6-phase	12-phase
		Half Wave	Full Wave CT	Full Wave Bridge	Star	Bridge	Star	Star
a	V_{DC}/V_m		0.6366	0.6366	0.8270	0.9549	0.9549	0.9886
b	V_{DC}/V_m'	—	—	—	—	1.654	—	—
c	V_{rms} of trans sec ph/V_{DC}	2.22	1.11	1.11	0.8550	0.4275	0.7405	0.7152
d	V_{rms} of trans sec ln-ln/V_{DC}	2.22	2.22	1.11	1.481	0.7405	1.481	1.430
e	PRV per diode	V_m	$2V_m$	V_m	$\sqrt{3}V_m$	$\sqrt{3}V_m'$	$2V_m$	$2V_m$
f	PRV per diode/V_{DC}	3.14	3.14	1.57	2.094	1.047	2.094	2.023
g	PRV per diode/V_{rms} of trans sec ph	1.41	2.82	1.41	2.449	2.449	2.828	2.828
h	PRV per diode/V_{rms} of trans sec ln-ln	1.41	1.41	1.41	1.414	1.414	1.414	1.41
i	I_{DC} per diode/I_{DC} Load	1.00	0.50	0.50	0.333	0.333	0.167	0.0833
j	peak I per diode/I_{DC} for R load	3.14	1.57	1.57	1.209	1.047	1.047	1.0115
k	peak I per diode/I_{DC} for L load	1.00	1.00	1.00	1.000	1.000	1.000	1.000
l	trans sec rms amps/I_{DC} load	1.57	0.707	1.00	0.5774	0.8165	0.4082	0.2887
m	trans sec VA/DC watts out	3.47	1.57	1.11	1.481	1.047	1.814	3.504
n	trans pri rms amps/I_{DC} load	1.57	1.00	1.00	0.4714	0.8165	0.5774	0.4082
o	trans pri VA/DC watts out	3.49	1.11	1.11	1.209	1.047	1.283	1.239
p	ave of pri & sec VA/DC watts out	3.49	1.34	1.11	1.345	1.047	1.548	2.371
q	utilization factor—primary		0.900	0.900	0.8267	0.9549	0.7797	0.8072
r	utilization factor—secondary		0.637	0.900	0.6752	0.9549	0.5513	0.2854
s	ripple frequency (F_1 is supply f)	F_1	$2F_1$	$2F_1$	$3F_1$	$6F_1$	$6F_1$	$12F_1$
t	fundamental V_{1max}/V_m				0.2067	0.0546	0.0546	0.04344
u	V_{1rms}/V_{DC}		0.472	0.472	0.1768	0.0404	0.0404	0.03107
v	2nd harmonic, V_{2max}/V_m				0.0473	0.0134	0.0134	0.003439
w	V_{2rms}/V_{DC}		0.095	0.095	0.0404	0.0099	0.0099	0.00246
x	3rd harmonic, V_{3max}/V_m				0.207	0.0069	0.0069	0.00153
y	V_{3rms}/V_{DC}		0.040	0.040	0.0177	0.0044	0.0044	0.00109
z	ripple factor, (%)	121	47	47	18.22	4.18	4.18	3.119
aa	line power factor		0.900	0.800	0.826	0.955	0.955	

Figure 8.2 Rectifier circuit constants.

* full-wave center tap (Figure 8.4)
* full-wave bridge (Figure 8.5)

Three-phase
* star (wye-connected secondary) (Figure 8.6)
  star (zig-zag-connected secondary)
* bridge (Figure 8.7)
  double-Y with interphase transformer

Six-phase (Figure 8.10)
* star (also called three-phase full wave or double-wye) with
  delta primary

> star with delta tertiary and wye primary
> fork
> open star or open double-wye
> bridge (or open double-wye)
> polygon

Twelve-phase

> star
> bridge (two three-phase bridges)

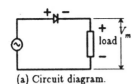

(a) Circuit diagram.

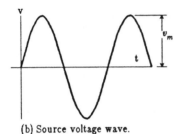

(b) Source voltage wave.

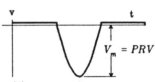

(d) Reverse voltage wave.

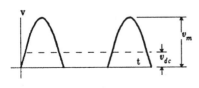

(c) Load voltage wave.

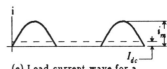

(e) Load current wave for a resistive load.

Figure 8.3 Single-phase half-wave rectifier.

8.2 SINGLE-PHASE RECTIFIER CIRCUITS

The simplest rectifier circuit is the single-phase half-wave rectifier of Figure 8.3a. For a sinusoidal input voltage (Figure 8.3b), the current will flow through the circuit whenever the voltage wave is positive so that the load voltage wave consists of the positive pulses of Figure 8.3c and the load

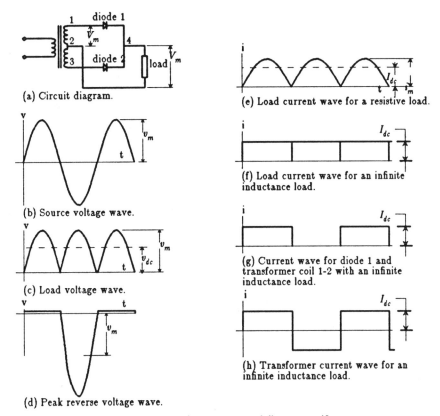

(a) Circuit diagram.

(b) Source voltage wave.

(c) Load voltage wave.

(d) Peak reverse voltage wave.

(e) Load current wave for a resistive load.

(f) Load current wave for an infinite inductance load.

(g) Current wave for diode 1 and transformer coil 1-2 with an infinite inductance load.

(h) Transformer current wave for an infinite inductance load.

Figure 8.4 Single-phase center-tap full-wave rectifier.

current for a pure resistance load is that of Figure 8.3e. If the load were a pure inductance, then the current wave would appear as a constant. The voltage across the diode is the difference between the supply voltage, Figure 8.3b, and the load voltage, Figure 8.3c, and is plotted as Figure 8.3d. The maximum value of this voltage is called the *peak-reverse-voltage* (PRV). Since the output or load voltage occurs half the time, the circuit is referred to as the *half-wave rectifier*.

There are two common circuits that provide a two-pulse output for a sinusoidal input and are therefore called *full-wave rectifier* circuits. The first requires two rectifiers and a transformer with a center tap, hence the name *center-tapped full-wave* rectifier (Figure 8.4a). The peak reverse voltage across each diode is the difference between the load voltage and the transformer voltage. For diode 1, the PRV occurs on every other positive output peak when diode 2 is conducting. At this same instant, transformer terminal 1 is at a negative peak. The voltage magnitude is the difference between a positive maximum at terminal 4 and a negative maximum at terminal 1, thus a PRV of $2V_m$. A more detailed proof is called for in Problem 8.12.

A second full-wave rectifier circuit is shown in Figure 8.5a. Since the four rectifiers are connected in a circuit similar to that of a Wheatstone resistance bridge, the name *full-wave bridge* rectifier has been given to this circuit. In this circuit two diodes conduct at any given instant. For the first positive output pulse of Figure 8.5c, diodes 1 and 2 conduct. For the second positive output pulse, diodes 3 and 4 conduct. The reverse voltage is divided between two diodes, hence PRV equals V_m.

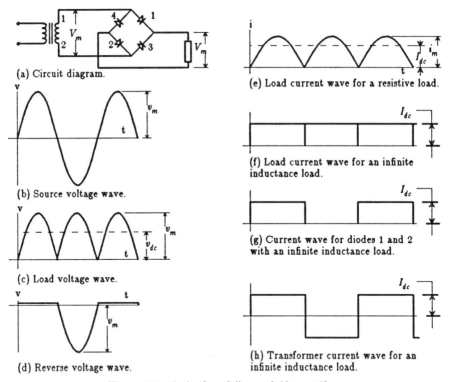

(a) Circuit diagram.

(b) Source voltage wave.

(c) Load voltage wave.

(d) Reverse voltage wave.

(e) Load current wave for a resistive load.

(f) Load current wave for an infinite inductance load.

(g) Current wave for diodes 1 and 2 with an infinite inductance load.

(h) Transformer current wave for an infinite inductance load.

Figure 8.5 Single-phase full-wave bridge rectifier.

The detailed analysis of the single-phase circuits is not presented in this text, but may be found in a basic text on electronic circuits. For larger amounts of dc power, the polyphase circuits are more efficient and require less filtering and are presented in more detail later in this chapter.

8.3 DEFINITION OF TERMS

Before considering the polyphase rectifier circuits, consider some definitions and terms. The first of these is the term *ripple factor*. However, the meaning of the terms *average* and *root-mean-square* must be understood first.

The average voltage wave $v(t)$ is

$$V_{ave} = V_{dc} = \frac{1}{T} \int_0^T v(t)dt \tag{8.1}$$

Two important average expressions are (1) that the average of a sine wave is zero, and (2) that the average of a sine wave between $0°$ and $180°$ is $2V_m/\pi = 0.637 V_m$. The root-mean-square (rms) of the voltage is defined as

$$V_{rms} = \left(\frac{1}{T} \int_0^T [v(t)]^2 dt \right)^{\frac{1}{2}} \tag{8.2}$$

The rms value for a sine wave is $V_{rms} = V_m/\sqrt{2}$. Another important rms expression is given for a repetitive wave expressed by a Fourier series. It is assumed that the reader is familiar with the Fourier series that is expressed for this writing in the form

$$v(t) = V_{dc} + V_{1m}\sin(\omega t + \theta_1) + V_{2m}\sin(2\omega t + \theta_2) \tag{8.3}$$
$$+ V_{3m}\sin(3\omega t + \theta_3) + \cdots$$

The rms value of (8.3) is

$$V_{rms} = \left[V_{dc}^2 + \left(\frac{V_{1m}}{\sqrt{2}}\right)^2 + \left(\frac{V_{2m}}{\sqrt{2}}\right)^2 + \left(\frac{V_{3m}}{\sqrt{2}}\right)^2 + \cdots \right]^{\frac{1}{2}} \tag{8.4}$$

Another form of (8.4) is

$$V_{rms} = \left[V_{dc}^2 + V_{1\ rms}^2 + V_{2\ rms}^2 + V_{3\ rms}^2 + \cdots \right]^{\frac{1}{2}} \tag{8.5}$$

The proof of (8.5) is left for Problem 8.8. For rectifier circuits there is a need for determining the rms of the ac terms with the dc term deleted. This is obtained by deleting the dc term of (8.5) so that

$$V_{(rms\ of\ the\ ac\ terms)} = \left[V_{1\ rms}^2 + V_{2\ rms}^2 + V_{3\ rms}^2 + \cdots \right]^{\frac{1}{2}} \tag{8.6}$$

Now the term *ripple-factor (RF)* [1] can be defined as

$$RF = \frac{\text{rms of the ac terms}}{\text{average value of the wave}}$$

$$= \frac{V_{rms \ of \ the \ ac \ terms}}{V_{dc}} \tag{8.7}$$

A term called the *transformer utilization factor* is also used in describing a rectifier circuit. The voltage wave for a rectifier input is a sine wave, but the current is non-sinusoidal. Since transformers are rated for use with sinusoidal voltages and sinusoidal current, the term transformer utilization factor is the ratio of the volt-amperes when used in a given rectifier circuit to the volt-amperes when used for sinusoidal waves. Thus,

$$UF = \frac{\text{VA of rectified circuit}}{\text{VA of equivalent sinusoidal wave}} \tag{8.8}$$

The wave shape of the current flowing through the transformer of a rectifier is obviously far from sinusoidal; therefore, the heat losses will be greater than under sinusoidal operation. Transformers, like other electrical equipment, are rated in terms of the output that they will deliver with a temperature rise not to exceed a predetermined safe value, assuming that they will be called upon to carry sine-wave currents. Ratings given on such a basis must therefore be altered for transformers used in rectifier circuits by multiplying them by a term known as the transformer utilization factor. Evaluation of this factor therefore requires that the heat losses be determined in the transformer when in rectifier service.

The impressed voltage across a transformer is ordinarily sinusoidal regardless of the type of load. Thus, the flux is nearly sinusoidal, and the iron losses may be considered independent of the wave shape of the load current. The winding losses, on the other hand, are directly dependent upon the current flowing through the windings and are therefore markedly affected by the wave shape. Under the assumption that there is infinite inductance in the load and the transformer secondary, the current for the three-phase half-wave rectifier of Figure 8.6 will be a rectangular pulse lasting for one-third of a cycle and having an amplitude equal to the direct load current (Figure 8.6h). Since the transformer is considered ideal, the primary current (Figure 8.6i) must be of the same shape except that no

---

[1] colloquial phrasing of (8.7) is

$$RF = \frac{garbage}{good}$$

direct component can be present since there is no source of direct voltage in the transformer primary circuit.

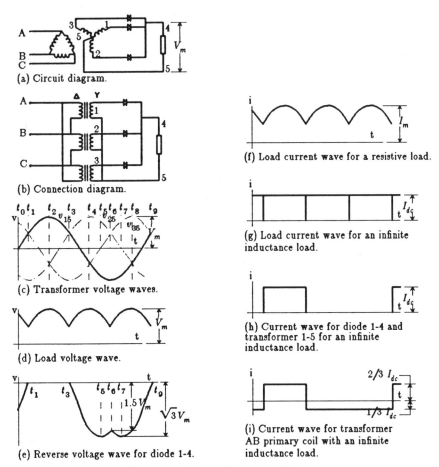

(a) Circuit diagram.

(b) Connection diagram.

(c) Transformer voltage waves.

(d) Load voltage wave.

(e) Reverse voltage wave for diode 1-4.

(f) Load current wave for a resistive load.

(g) Load current wave for an infinite inductance load.

(h) Current wave for diode 1-4 and transformer 1-5 for an infinite inductance load.

(i) Current wave for transformer AB primary coil with an infinite inductance load.

Figure 8.6 Three-phase star rectifier.

8.4 POLYPHASE RECTIFIER CIRCUITS

The possible polyphase circuits are numerous. Three of the most common will be shown in this chapter with the detailed computation presented for the three-phase bridge.

A. Three-Phase Star

The simplest three-phase rectifier has been called by such names as *three-phase star* and *three-phase half-wave* rectifier circuit and is shown in Figure 8.6. It should be noted that the transformer primaries could be connected wye or delta. Any of the three diodes will conduct when the anode is more positive than the cathode. Thus, between the times t_1 and t_3 of Figure 8.6c, the voltage potential at the anode of diode 1 is more positive than the load voltage at terminal 4 and diode 1 is the conducting diode. From time t_3 to t_6, diode 2 is the conducting diode. Finally, from time t_6 to t_9, diode 3 is the conducting diode. The potential across the load is that shown in Figure 8.6d.

The reverse voltage across each diode has a shape similar to that shown in Figure 8.6e. For diode 1-4 the reverse voltage is the maximum difference between the solid curve and the dashed curves of Figure 8.6c. When the voltage at the anode of diode 1 is a negative maximum, the reverse-voltage is $1.5\ V_m'$. Often this value is assumed to be the maximum or peak-reverse-voltage (PRV)[2]. However, at times t_5 and t_7 the lower curve is $(\sqrt{3}/2)\ V_m$, and the upper curve is $(\sqrt{3}/2)\ V_m$. Thus, the PRV is the maximum difference between the curves and is $\sqrt{3}\ V_m$.

B. Three-phase Bridge Rectifier.

The circuit shown in Figure 8.7 is referred to as a *bridge rectifier* circuit. Other names that are sometimes used are "three-phase full wave" and "six-phase double-way." The rectifiers conduct in pairs to provide a path for current through the transformer from the upper to the lower output terminal. The conduction periods of the rectifiers within the upper and lower output terminal sets do not coincide in time, however. Each rectifier conducts for 120 degrees per cycle corresponding to the lobe along the upper and lower envelope of the voltage waves, as is indicated in Figure 8.7b. With *abc* phase sequence, the rectifiers conduct in pairs in the order 1-6, 3-6, 3-2, 5-2, 5-4, 1-4, 1-6, and so on. Since each rectifier conducts one-third of the time, its required average current rating is one-third of the average output current.

Each transformer secondary winding carries the current conducted by two rectifiers. These currents flow in opposite directions through the

---

[2]The peak-reverse-voltage (PRV) is also called peak-inverse-voltage (PIV).

windings, hence the average winding current is zero. Direct current magnetization of the core, and consequently excessive exciting current, are thereby eliminated. Furthermore, since the winding carries current in one direction or the other for two-thirds of the time, the secondary utilization factor for this connection is relatively large.

Note that V_m is the peak value for the output voltage wave of Figure 8.7c. This is also the peak value of the transformer secondary line-to-line voltage. In contrast, V_m' is the peak value of the voltage waves for Figure 8.7b. The value, V_m', is also the peak value of the transformer secondary line-to-neutral voltage.

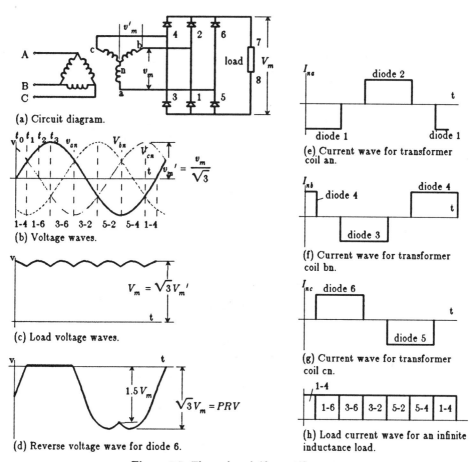

(a) Circuit diagram.

(b) Voltage waves.

(c) Load voltage waves.

(d) Reverse voltage wave for diode 6.

(e) Current wave for transformer coil an.

(f) Current wave for transformer coil bn.

(g) Current wave for transformer coil cn.

(h) Load current wave for an infinite inductance load.

Figure 8.7 Three-phase bridge rectifier.

Example 8.1 Verify the values in Figure 8.2 for the bridge rectifier circuit.

Solution

    Begin by calculating the Fourier series terms for a six-pulse curve shown in Figure 8.8. Equation (8.9) is the half-range expansion expression for the Fourier series terms as explained in many mathematics texts [O.4, p.363].

$$a_n = \frac{2}{p} \int_0^p f(\omega t) \cos\left(\frac{n\pi\omega t}{p}\right) d(\omega t) \qquad (8.9)$$

where

$$f(\omega t) = V_m \cos\omega t \quad \text{and} \quad p = \pi/6$$

Thus the general expression for the terms is

$$a_n = \left(\frac{2V_m}{\pi/6}\right) \int_0^{\pi/6} (\cos\omega t)\left[\left(\frac{n\pi}{\pi/6}\right)\omega t\right] d(\omega t) \qquad (8.10)$$

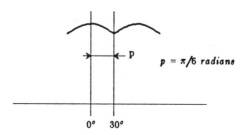

$p = \pi/6$ *radians*

$0°\quad 30°$

Figure 8.8 Rectified wave for Example 8.1.

Perform the integration of (8.10) to obtain the expression

$$a_n = \frac{12V_m}{\pi}\left[\frac{\sin(1-6n)\omega t}{2(1-6n)} + \frac{\sin(1+6n)\omega t}{2(1+6n)}\right]_0^{\omega t = \pi/6} \qquad (8.11)$$

The steady state term of the series is

$$a_0 = \frac{6V_m}{\pi}\left[\sin(\pi/6) + \sin(\pi/6)\right] = \frac{6V_m}{\pi}\left(\frac{1}{2} + \frac{1}{2}\right) = \frac{6V_m}{\pi}$$

Thus the value for the dc voltage is

$$V_{dc} = \frac{a_o}{2} = \frac{3V_m}{\pi} = 0.954\ 930\ V_m$$

or, from Figure 8.7

$$V_{dc} = \frac{3\sqrt{3}\,V_m{}'}{\pi} = 1.653\ 99\,V_m{}' \tag{8.12}$$

The fundamental term is obtained from (8.11) where $n = 1$

$$a_1 = \frac{6V_m}{\pi}\left[\frac{\sin(-5\pi/6)}{-5} + \frac{\sin(7\pi/6)}{7}\right] \tag{8.13}$$

$$= \frac{6V_m}{35\pi} = 0.054\ 567\ V_m$$

The second harmonic term where $n = 2$ is

$$a_2 = \frac{6V_m}{\pi}\left[\frac{\sin(-11\pi/6)}{-11} + \frac{\sin(13\pi/6)}{13}\right] \tag{8.14}$$

$$= \frac{-6V_m}{143\pi} = -0.013\ 356\ V_m$$

and finally, the third harmonic term where $n = 3$ is

$$a_3 = \frac{6V_m}{\pi}\left[\frac{\sin(-17\pi/6)}{-17} + \frac{\sin(19\pi/6)}{19}\right] \tag{8.15}$$

$$= \frac{6V_m}{323\pi} = 5.912\ 88 \times 10^{-3}\ V_m$$

Combine the values obtained in (8.12) through (8.15) to determine the instantaneous expression for the voltage

$$v(t) = V_m\left[\frac{3}{\pi} + \left(\frac{6}{35\pi}\right)\sin 6\omega t - \left(\frac{6}{143\pi}\right)\sin 12\omega t\right. \tag{8.16}$$

$$\left. + \left(\frac{6}{323\pi}\right)\sin 18\omega t \cdots\right]$$

With the Fourier series expression of (8.16), most of the constants in rows a through z of Figure 8.2 may be determined for the three-phase bridge circuit.

a. From (8.12), $\dfrac{V_{dc}}{V_m} = \dfrac{3}{\pi} = 0.954\ 930$

b. $\dfrac{V_{dc}}{V_m'} = \dfrac{3\sqrt{3}}{\pi} = 1.653\ 99$ (also see (8.12))

c. $\dfrac{V_{rms}}{V_{dc}} = \dfrac{(V_m'/\sqrt{2})}{V_{dc}} = \left(\dfrac{1}{\sqrt{2}}\right)\left(\dfrac{\pi}{3\sqrt{3}}\right) = \dfrac{\pi}{3\sqrt{6}} = 0.427\ 517$

d. $\dfrac{V_{rms}\,line}{V_{dc}} = \sqrt{3}\left(\dfrac{V_{rms}}{V_{dc}}\right) = \sqrt{3}\left(\dfrac{\pi\sqrt{3}}{6}\right) = \dfrac{\pi}{3\sqrt{2}} = 0.740\ 480$

e. The PRV is the maximum value of the difference between the voltage of curve 2 and curve 3 for the time period between time t_4 and t_6 of Figure 8.7b.

$$v_{23} = V_m\left[\sin\omega t - \sin(\omega t - 120°)\right]$$
$$= V_m(\sin\omega t - \sin\omega t\,\cos120° + \cos\omega t\,\sin120°)$$
$$= V_m'\left[\left(\dfrac{\sqrt{3}}{2}\right)\sin\omega t + \left(\dfrac{\sqrt{3}}{2}\right)\cos\omega t\right]$$
$$= \sqrt{3}\,V_m'\sin(\omega t + 30°)$$
$$\text{PRV} = \text{max of } \left|\sqrt{3}\,V_m'\sin(\omega t + 30°)\right| = \sqrt{3}\,V_m' \tag{8.17}$$

also

$$\text{PRV} = V_m = \sqrt{3}\,V_m' \tag{8.18}$$

f.

$$\dfrac{\text{PRV per diode}}{V_{dc}} = \sqrt{3}\left(\dfrac{V_m'}{V_{dc}}\right) = \sqrt{3}\left(\dfrac{\pi\sqrt{3}}{3}\right) \tag{8.19}$$

$$= \dfrac{\pi}{3} = 1.047\ 20$$

g.

$$\dfrac{\text{PRV}}{V_{rms}\ per\ leg} = \dfrac{\sqrt{3}\,V_m'}{V_{rms}} = \dfrac{\sqrt{3}\,V_m'}{V_m'/\sqrt{2}} = \sqrt{6} = 2.449\ 49 \tag{8.20}$$

h.

$$\frac{PRV}{V_{rms}\,line} = \frac{PRV\,/\,V_{rms}per\;leg}{\sqrt{3}} = \frac{\sqrt{6}}{\sqrt{3}} = \sqrt{2} = 1.414\;21 \quad (8.21)$$

i.

$$\frac{I_{dc\;per\;diode}}{I_{dc\;load}} = \frac{1}{3} \quad\quad (8.22)$$

j. This is the reciprocal of row (a) of Figure 8.2. Thus

$$\frac{V_m}{V_{dc}} = \frac{\pi}{3} = 1.047\;20 \quad\quad (8.23)$$

k. For a pure inductive load the current is constant. Therefore,

$$V_m = V_{dc} \quad \text{and} \quad \frac{V_m}{V_{dc}} = 1.0$$

l. A graphical integration by observing the area under the I_{dc}^2 curve shown in Figure 8.9 provides

$$I_{rms} = \left\{ \frac{1}{2\pi} \left[\left(\frac{2\pi}{3}\right)I_{dc}^2 + \left(\frac{2\pi}{3}\right)I_{dc}^2 \right] \right\}^{\frac{1}{2}}$$

$$= I_{dc}^2 \left(\frac{4\pi/3}{2\pi}\right)^{\frac{1}{2}} = \sqrt{2/3}\;I_{dc} = 0.816\;497 I_{dc}$$

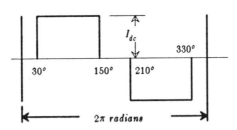

Figure 8.9 Transformer secondary current.

m. The transformer secondary winding loss, P_{2r}, is

$$P_{2r} = \frac{1}{T} \left[I_{dc}^2 R_2 \left(\frac{T}{3}\right) + I_{dc}^2 R_2 \left(\frac{T}{3}\right) \right] = \frac{2I_{dc}^2 R_2}{3}$$

where T is the period and R_2 is the secondary coil resistance. A standard transformer rating is $P_{2n} = I_2^2 R_2$. But

$$P_{2n} = P_{2r} \quad \text{or} \quad I_2^2 R_2 = \frac{2 I_{dc}^2 R_2}{3}$$

Then

$$I_2 = \sqrt{2/3} \, I_{dc} = 0.816\ 497 \, I_{dc}$$

$$E_2 = 0.4275 \, E_{dc} \quad \text{from row (c)}$$

$$E_2 I_2 = \left(0.4275 E_{dc}\right)\left(0.816\ 497 \, I_{dc}\right)$$

$$= \left(\frac{\pi}{3\sqrt{6}}\right) E_{dc} \left(\sqrt{2/3} \, I_o\right) = \left(\frac{\pi}{9}\right) E_{dc} I_{dc}$$

$$= 0.349\ 066 \, E_{dc} I_{dc}$$

$$(E_2 I_2) \text{ for three-phase} = 3(E_2 I_2) = 3 \left(\frac{\pi}{9}\right) E_{dc} I_{dc}$$

$$= \left(\frac{\pi}{3}\right) E_{dc} I_{dc} = \left(\frac{\pi}{3}\right) P_{dc} = 1.047\ 20 \, P_{dc}$$

n. The (transformer primary A)$/I_{dc}$ load is the same as the secondary rms amperes obtained in row (l).

o. Same as secondary — see row (m).

p Same as rows (o) and (m).

q. $\dfrac{1}{\text{row (m)}} = \dfrac{3}{\pi} = 0.954\ 930$

r. Same as row (q).

s. The output ripple frequency is 6 times the supply frequency as noted from the output voltage shown in Figure 8.7c.

t. $\dfrac{V_{1\,max}}{V_m} = \dfrac{6}{35\pi} = 0.054\ 567\ 4$

u. $\dfrac{V_{1\,rms}}{V_{dc}} = \dfrac{\left(\dfrac{V_{1\,max}}{\sqrt{2}}\right)}{V_{dc}}$

from row (a), $\quad \dfrac{V_{dc}}{V_m} = \dfrac{3}{\pi}$

Thus,

$$\frac{V_{1\,rms}}{V_{dc}} = \frac{\left(\dfrac{V_{1\,max}}{\sqrt{2}}\right)}{\left(\dfrac{3V_m}{\pi}\right)} = \left(\frac{V_{1\,max}}{V_m}\right)\left(\frac{\pi}{3\sqrt{2}}\right)$$

but $\dfrac{V_{1\,max}}{V_m} = \dfrac{6}{35\pi}$ from row (t). Finally,

$$\frac{V_{1\,rms}}{V_{dc}} = \left(\frac{6}{35\pi}\right)\left(\frac{\pi}{3\sqrt{2}}\right) = \frac{\sqrt{2}}{35} = 0.040\ 406$$

v. $\quad \dfrac{V_{2\,max}}{V_m} = \dfrac{a_2}{V_m} = \dfrac{6}{143\pi} = 0.013\ 356$

w. $\quad \dfrac{V_{2\,rms}}{V_{dc}} = \dfrac{V_{2\,max}/\sqrt{2}}{V_{dc}} = \dfrac{V_{2\,max}/\sqrt{2}}{3V_m/\pi}$

$$= \frac{\sqrt{2}}{143} = 0.009\ 889\ 6$$

x. $\quad \dfrac{V_{3\,max}}{V_m} = \dfrac{a_3}{V_m} = \dfrac{6}{323\pi} = 5.91288 \times 10^{-3}$

y. $\quad \dfrac{V_{3\,rms}}{V_{dc}} = \dfrac{V_{3\,max}/\sqrt{2}}{V_{dc}} = \dfrac{V_{3\,max}/\sqrt{2}}{3V_m/\pi}$

$$= \frac{\sqrt{2}}{323} = 0.004\ 378\ 37$$

z. ripple factor $\approx \dfrac{\left(V_{1\,rms}^2 + V_{2\,rms}^2 + V_{3\,rms}^2\right)^{1/2}}{V_{dc}}$

$$= \frac{\left(V_{dc}^2\left((0.040\ 406)^2 + (0.009\ 889\ 6)^2 + (0.004\ 378\ 37)^2\right)\right)^{\frac{1}{2}}}{V_{dc}}$$

$$= 0.041\ 829$$

Example 8.2. Specify the ratings for the transformers and diodes used in a three-phase bridge circuit that supplies a 375 kW, 300 V, dc load. The source is 4160 V, three-phase, and 60 Hz. Consider a delta-wye transformer connection.

Solution

Use the circuit of Figure 8.7 and the constants (a-z) of Figure 8.2.

Load

$$I_{(rated\ load)} = I_{dc} = \frac{375,000}{300} = 1250 \text{ A}$$

Transformer
Row (c) of Figure 8.2:

$$V_{(rms\ per\ leg)} = 0.4275\ V_{dc} = 0.4275(300) = 128 \text{ V}$$

Transformer turns ratio = 4160/128 = 32.4 for a delta-wye connection Row (p) of Figure 8.2: kVA = 1.047 (kw) = 1.047(375) = 393 kVA. Therefore, the requirement is (a) one 393 kVA, 4160:222 V, three-phase transformer or (b) three 131 kVA, 4160:128 V single-phase transformers.

Diode
Row (f): PRV = 1.047(300) = 314 V
Row (i): $I_{(dc\ per\ diode)}$ = 0.3333(1250) = 417 A
Row (j): $I_{(peak\ for\ resistive\ load)}$ = 1.047(1250) = 1309 A
Row (k): $I_{(peak\ for\ inductive\ load)}$ = 1.0(1250) = 1250 A

Therefore, the diode shall be rated at least
    peak reverse voltage = 314 V
    peak current = 1309 A
    average current = 417 A

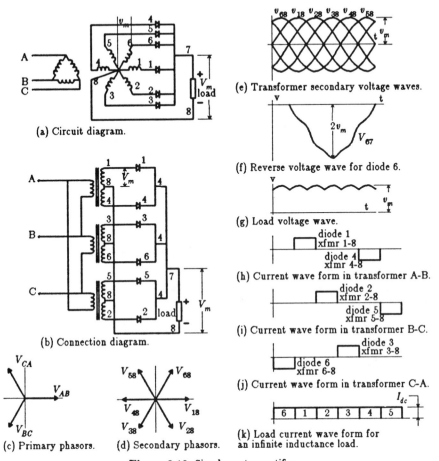

(a) Circuit diagram.

(b) Connection diagram.

(c) Primary phasors.

(d) Secondary phasors.

(e) Transformer secondary voltage waves.

(f) Reverse voltage wave for diode 6.

(g) Load voltage wave.

(h) Current wave form in transformer A-B.

(i) Current wave form in transformer B-C.

(j) Current wave form in transformer C-A.

(k) Load current wave form for an infinite inductance load.

Figure 8.10 Six-phase star rectifier.

PROBLEMS

8.1 What is the expected meter indication when a d'Arsonval meter is used to measure the voltage described by Figures 8.11a and 8.11b?

8.2 Repeat Problem 8.1 for an iron-vane voltmeter.

8.3 Repeat Problem 8.1 when the voltage source is modified through a half-wave rectifier.

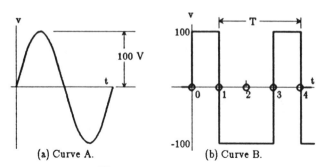

(a) Curve A.   (b) Curve B.

Figure 8.11 Signal waves.

8.4 Repeat Problem 8.2 when the voltage source is modified through a half-wave rectifier.

8.5 What is the expected meter indication for a VTVM voltmeter for the voltage described in Figures 8.11a and 8.11b?

8.6 What is the expected meter indication for a peak indicating voltmeter for the voltage described in Figure 8.11a and 8.11b?

8.7 Derive the expression for the average of a half-sine wave in terms of the maximum value.

8.8 Prove Equation 8.5.

8.9 What is the effective (or rms) value of the voltage wave described by the expression

$$v(t) = 50 + 83 \sin (50t + 25^\circ) + 50 \sin (100t + 15^\circ) - 25 \sin (150t - 30^\circ)$$

8.10 What is the effective (or rms) value of the ac terms of the voltage wave of Problem 8.9?

8.11 What is the ripple factor for the voltage wave of Problem 8.9?

8.12 An important characteristic of a rectifier circuit is the peak-reverse-voltage across each diode. (a) Determine the PRV for a half-wave rectifier in terms of the dc or average value of the output voltage. (b) Repeat *a* for a full-wave center tap circuit. (c) Repeat *a* for a full-wave bridge circuit. Determine the PRV for parts (a) through (c) in terms of the peak or maximum of the ac input wave.

8.13 Diodes are available that can withstand a PRV of 25 V each and an average current of 100 mA in the forward direction. It is desired to use some of these diodes for a single-phase bridge rectifier that can be used with a 120 V ac source and that will be capable of delivering an average current of 200 mA to a pure resistance load. Diodes may be connected in series and/or parallel to obtain higher voltage and/or current ratings for the various arms of the bridge. What is the minimum number of diodes that will be needed if none of the ratings are to be exceeded, and how should those diodes be arranged?

8.14 A three-phase star rectifier is used to convert 480 V, three-phase, 60 Hz to dc for supplying a 200 kW, 250 V, dc load. Use a delta primary transformer connection. (a) Sketch the circuit diagram. (b) Specify the transformer turns ratio. (c) Specify the diode PRV. (d) What is the average diode current? (e) What is the peak diode current? (f) Specify the total transformer volt-ampere rating.

8.15 Repeat Problem 8.14 for a three-phase bridge system with a delta-wye transformer connection.

8.16 Repeat Problem 8.14 for a six-phase star with a delta primary.

8.17 Graphically and analytically verify that the PRV for the center-tapped full-wave single-phase rectifier is $2V_m$ as shown in Figure 8.4.

8.18 Given three single-phase transformers each rated 100 kVA, 480:240/120 V. These transformers supply energy to a three-phase star rectifier. What is (a) the dc load voltage, and (b) the maximum dc wattage without overloading the transformers? Consider the transformers connected delta-wye with the low voltage coils in series for a secondary voltage of 240 V per phase.

8.19 Repeat Problem 8.18 for a (a) three-phase bridge rectifier, (b) six-phase star rectifier with 120 V per transformer secondary.

8.20 Given three laboratory transformers each rated 208:104 V, 0.40:0.80 A and a set of diodes rated 600 V PRV, 1 A average, and 30 A peak. Connect these transformers in a three-phase star rectifier circuit. (a) What is the largest dc load current that does not exceed the transformer rating? (b) What is the largest dc load current that does not exceed the diode rating? (c) What is the largest dc load current that will not overload the circuit elements?

8.21 Repeat Problem 8.20 for a three-phase bridge circuit.

9

dc MACHINES

The dc machine as a generator was the first device used to provide a significant amount of electrical energy. The dc motor was also the first electrical device to provide rotating mechanical energy. Recently, however, polyphase rectifier systems are replacing dc generators in industrial plants.

The first dc generator (or dynamo as it was then called) was a copper disk rotated between the poles of a horseshoe magnet. This was an experiment described by Michael Faraday, an English chemist and physicist, in 1831. Joseph Henry, an American physicist, invented a rocker arm motor the same year.

Antonio Pacinothi and Sir William Siemens introduced the commutator in 1860. Zenobe Theophile Gramme invented the first efficient industrial dc generator, known as the Gramme machine, in 1869. By accident, it was found that this machine would also work as a motor. This machine was displayed at the Philadelphia Centennial Exhibition in 1876. It was seen independently by Elihu Thompson and Charles Brush, who developed improved dynamos for their electric arc lamp systems.

Edison installed the "long-waisted Mary Anns," developed by Frances Upton for the Pearl Street installation in New York, in 1882. These were driven by a double-acting steam engine. Many dc installations followed; however, these were limited to short distances because of the line voltage drop. The invention of the ac transformer provided the means for an economical change between high and low voltages for the economical transmission of large quantities of electrical energy. Thus, there are few systems, except in manufacturing plants, where special dc motors, electrolytic processes, or special computers are required.

9.1 THEORY

There is much that could be discussed about the details of dc machine design. This includes the type of core and yoke metal, magnetic field plots, lap and wave windings, number of conductors per slot, armature laminations, etc. For most engineers, these topics are of as much real concern as the amount of doping of a silicon wafer for a transistor or the thickness of the cylinder walls in a gasoline engine. This chapter deals with the applications of the dc machine and not with detailed design procedures.

A voltage is generated by moving an electric conductor in a magnetic field. The vector equation is

$$\vec{V}_g = l\,(\,\vec{u} \times \vec{B}\,) \tag{9.1}$$

where $\vec{V}_g$ is the induced voltage
l is the length of the conductor in meters
$\vec{u}$ is the velocity of motion in meters/second
$\vec{B}$ is the magnetic flux density in teslas

For a dc machine, (9.1) is expressed more usefully as

$$V_g = \left(\frac{PZ}{2\pi a}\right)\phi\omega = K\phi\omega \tag{9.2}$$

where P is the number of magnetic poles
Z is the number of conductors
a is the number of parallel paths
ω is the speed of rotation in radians/second
$K = \dfrac{PZ}{2\pi a}$
ϕ is the magnetic flux in webers

Another important equation relates the motor torque to the magnetic field and the armature current. But first consider the force on a simple conductor in a magnetic field

$$\vec{F} = i\vec{l} \times \vec{B} \qquad (9.3)$$

where $\vec{F}$ = force in newtons
$\vec{i}$ = conductor current in amperes
l = length of conductor in meters
$\vec{B}$ = flux density in the teslas

Also,

$$\vec{T} = \vec{r} \times \vec{F} \qquad (9.4)$$

where $\vec{r}$ = radius in meters
$\vec{F}$ = force in newtons calculated from (9.3)
$\vec{T}$ = torque in newton-meters

For a dc machine where the vectors of (9.4) are mutually perpendicular, the equation may be written as

$$T = \frac{PZ}{2\pi a} \phi i_a = K\phi i_a \qquad (9.5)$$

where p = number of magnetic poles
Z = number of conductors
ϕ = flux in webers
i_a = armature current in amperes
$K = \dfrac{pZ}{2\pi a}$

Having outlined the introductory equations, note that the dc machine is bidirectional. It can be used equally well as a motor or generator, although speed, application, and other mechanical features may dictate special construction. Now, consider the machine construction.

9.2 CONSTRUCTION

The dc machine consists of three basic parts: (1) the field and yoke, (2) the armature, and (3) the commutator (connected to the armature). Various parts of the machine are shown in Figures 9.1-9.3. Figure 9.2 shows the field assembly with the main pole and winding and the yoke of cast iron that provides the mechanical strength and a path for magnetic flux. The commutating poles (also called interpoles) allow for reversing operations particularly on large machines.

The armature shown in Figure 9.3 consists of (1) a shaft, (2) a laminated iron structure on the shaft, (3) windings, and (4) a commutator.

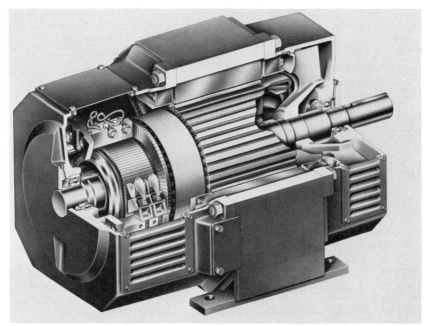

Figure 9.1 Cut-away view of a dc machine. *(Courtesy of General Electric Company)*

Figure 9.2 Frame of a dc machine showing the yoke, poles, inter-poles and windings. *(Courtesy of General Electric Company)*

Figure 9.3 Armature of a dc machine showing coils and commutator. *(Courtesy of General Electric Company)*

The windings may be preformed coils or may be randomly wound wiring. The random winding is generally found on fractional horsepower sizes.

The commutator assembly in conjunction with the brushes is essentially a mechanical rectifier. For a generator, the commutator keeps the external electric current flowing in one direction. For a motor application, the commutator keeps the electric current flowing in the proper direction to provide a unidirectional torque. All commutator segments must be electrically isolated from each other and from the shaft. It is important to note that the commutator assembly is an expensive part of the machine. The brushes that rub against the commutator require maintenance and must be replaced when worn. Any oil applied to the commutator by a well-meaning but ill-informed maintenance man will cause poor brush contact and short out the commutator segments, and the arcing will often destroy the insulation. Also, too much grease in the bearing may eventually work onto the commutator assembly.

The details of armature coil connections are rather involved and not absolutely required in understanding the application of dc machines. It should be noted that each coil may consist of more than one conductor.

Even though the armature assembly consists of many parts, the dc small fractional horsepower machine can run faster than the squirrel-cage induction machine and is better suited for high speed applications such as vacuum cleaners and small high-speed fans.

9.3 COMMUTATOR ACTION

The flux distribution in the air gap of an elementary two-pole dc machine (Figure 9.4) is produced by the field and is shown in Figure 9.5a. A single coil moved through this flux will generate an alternating voltage. The commutator mechanically reverses the voltage direction to produce the voltage wave shown in Figure 9.5b. Electronic rectifiers called diodes could be used to perform this function as is shown for the automobile alternator described in Chapter 11. As more coils are added to the commutator-type dc machine, the voltage more nearly approaches a constant potential, as shown in Figure 9.6.

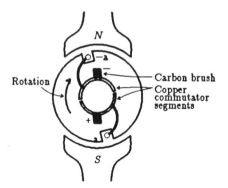

Figure 9.4 Elementary dc machine with commutator.

When the dc machine is used as a motor, a torque results, as shown in Figure 9.7. At position *a*, there is a small clockwise torque. At position *b*, the torque is a maximum. Then, at position *c*, the torque is zero. At position *c*, the current must be reversed through the conductor in order to keep the torque in a clockwise direction. The commutator performs this function. The torque produced by each coil is zero two times during each revolution. If the motor is stopped and restart attempted at a zero torque position, the motor will fail to move. This problem is eliminated by using multiple coils, as shown in Figure 9.8. In this assembly, the torque is always positive. As more commutator segments are added, the torque

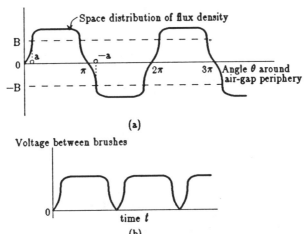

(a)

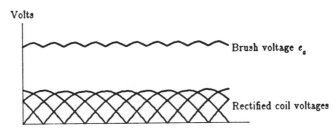

(b)

Figure 9.5 (a) Space distribution of air-gap flux density in an elementary dc machine and (b) waveform of voltage between brushes.

Figure 9.6 Rectified coil voltages and resultant voltage between brushes in a dc machine.

approaches a constant magnitude. The above is a simplified description. The designer must consider additional features as more coils and poles are added.

9.4 dc MACHINE CONNECTIONS

The advantages of dc machines arise from the variety of operating characteristics obtained from the various methods of connecting the field windings. Before explaining the types of connections, consider the field winding construction. The field may be wound with one or more coils of wire. These coils may consist of many turns of fine wire for small currents or a

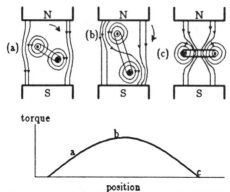

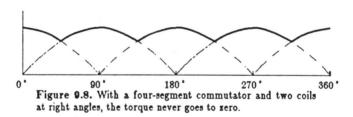

Figure 9.7 A current-carrying loop in a magnetic field builds up torque through one-half revolution. To obtain a complete revolution the current in the loop must be reversed at position (c) and again just ahead of (a). Only a portion of the field flux is shown. It is assumed to be uniformly distributed over the pole faces.

Figure 9.8. With a four-segment commutator and two coils at right angles, the torque never goes to zero.

few turns of heavy wire for large currents. The coil consisting of many turns of fine wire is designed to be connected across (or in shunt with) the machine terminals, hence the name *shunt coil*. The coil consisting of a few turns of heavy wire is designed to be connected in series with the armature and is called the *series coil*. A machine with only a shunt coil is called a *shunt machine*. A machine with only a series coil is called a *series machine*. When both coils are wound on each pole, the machine is said to be a *compound machine*. Some specialty motors, particularly for control applications, may have four or more coils wound on each pole. Figure 9.9 shows a single pole with a series and a shunt winding. The polarity of the compound machine coil connection is important since the flux produced in each winding may oppose or aid that of the other winding.

If the fluxes of the two windings are additive, the connection is said to be cumulative. If the fluxes of the two windings are subtractive, the connection is differential.

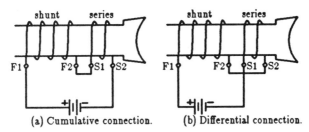

(a) Cumulative connection.   (b) Differential connection.

Figure 9.9 Magnetic effect on dc compound machine field connections.

9.5 MAGNETIZATION CURVE AND THE EQUIVALENT CIRCUIT

A simple mathematical model of a dc machine is the Thevenin equivalent, which contains an impedance in series with a voltage source, as shown in Figure 9.10. To the Thevenin equivalent add the field circuits and the prime mover or mechanical load. The mechanical features are generally shown dotted. The field and armature inductances are not shown for the steady state dc model since the reactances are zero for steady state dc operation. However, the inductances should be shown for the transient solution.

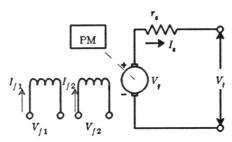

Figure 9.10 dc machine equivalent circuit.

The dc machine problem would be a simple linear dc circuit problem if it were not for the magnetic saturation of the field circuit. When saturation is considered, the analysis is more easily handled as a graphical-analytical procedure. The magnetization curve is determined experimentally by measuring the armature terminal voltage with no load as the armature is driven at a constant speed and as the field current is gradually increased. Typical magnetization curves are shown in Figure 9.11. It becomes apparent that this is a multi-dimensional problem. Fortunately, one magnetization curve provides sufficient information for determining the magnetization curves for various speeds. This is done by using (9.2) for two conditions so that

$$\frac{V_{g2}}{V_{g1}} = \frac{K_2\phi_2\omega_2}{K_1\phi_1\omega_1} \tag{9.6}$$

Since $K_2 = K_1$ for a given machine, and if ϕ_1 is selected equal to ϕ_2, as is the case when the field current is held constant, then

$$\frac{V_{g2}}{V_{g1}} = \frac{\omega_2}{\omega_1} = \frac{n_2}{n_1} \tag{9.7}$$

Where ω is the angular velocity in radians per second and n is the shaft speed in revolutions per minute. The procedure is shown in Examples 9.1 and 9.2.

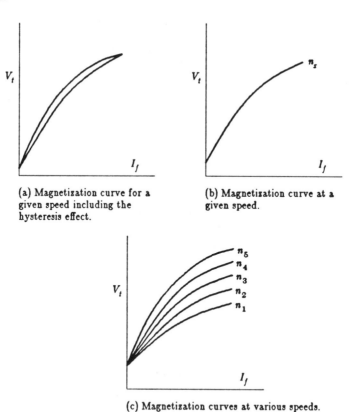

(a) Magnetization curve for a given speed including the hysteresis effect.

(b) Magnetization curve at a given speed.

(c) Magnetization curves at various speeds.

Figure 9.11 Magnetization curves for a dc machine.

Example 9.1

A 50 kW, 250 V, 1750 rev/min dc shunt machine has an armature resistance of 0.1 Ω and a shunt field resistance of 20 Ω. The data for a magnetization curve at 1750 rev/min are given below and in Figure 9.12.

I_f	0	1	2	3	4	5	6	7	8	9	10
V_t	8	28	60	110	158	208	237	255	268	280	287

(a) Determine the magnetization curve for 2100 rev/min.

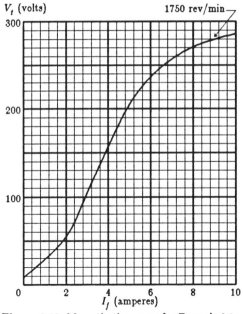

Figure 9.12 Magnetization curve for Example 9.1.

Solution

Use (9.7) to determine a ratio between voltages so that

$$\frac{V_{g2}}{V_{g1}} = \frac{n_2}{n_1}$$

and

$$V_{g2} = \left(\frac{2100}{1750}\right)V_{g1} = 1.20\,V_{g1}$$

(b) Next determine values of V_{g1} for various values of field current and solve for the related value of V_{g2}. Two V_{g2} points on the 2100 rev/min curve are 132 V for 3 A and 336 V for 9 A. The full curve is shown in Figure 9.13.

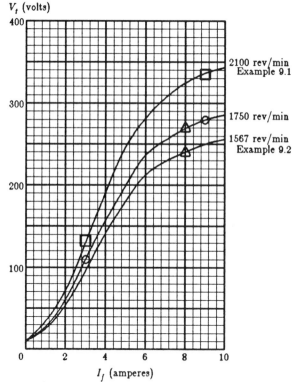

Figure 9.13 Examples 9.1 and 9.2.

Example 9.2

The machine of Example 9.1 is connected as a separately excited generator, and the shunt field is adjusted to 8.0 A. How fast must the machine armature be rotated for a no-load terminal voltage of 240 V?

Solution

The generated voltage V_{g1} is 268 V for a speed of 1750 rev/min. By using (9.7), the speed for a generated voltage V_{g2} is

$$N_2 = N_1 \frac{V_{g2}}{V_{g1}} = 1750 \left(\frac{240}{268}\right) = 1567 \ \text{rev/min}$$

This point is shown as one point on the 1567 rev/min curve on Figure 9.13.

9.6 LOAD CHARACTERISTICS FOR A dc GENERATOR

The generator characteristics are different for each of the several types of field connections. In fact, it took 25 years from the time of Faraday in 1831 until Siemens in 1866 to come up with the self-excited machine. During that period, only permanent magnets were used for the machine field. Possible connections are shown in Figure 9.14, and the related load curves are shown in Figures 9.15 and 9.20. A list of the normal connections includes the following:

> Permanent magnetic field
> Separately excited field
> Self-excited field
>   — Shunt
>   — Series
>   — Compound
>   — Cumulative
>   — Over compound
>   — Flat compound
>   — Under compound
>   — Differential

A. Permanent Magnet Generators

One important application for the permanent magnet generator is the dc tachometer. Generally, these are small machines that operate in the linear portion of the magnetization curve and provide a voltage output that is proportional to the speed. Usually the output is supplied to a meter, an electronic amplifier, or the field of a rotational amplifier.

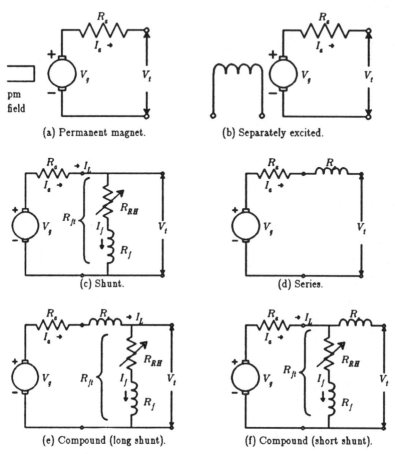

Figure 9.14 shows six field connection diagrams labeled:

(a) Permanent magnet.

(b) Separately excited.

(c) Shunt.

(d) Series.

(e) Compound (long shunt).

(f) Compound (short shunt).

Figure 9.14 Field connections for dc machines.

B. Load Characteristic for a Separately Excited Generator

An important generator characteristic is the effect that varying the load current has on the terminal voltage. For a separately excited connection, the terminal voltage, V_t, would obviously be less than the generated voltage, V_g, by the IR drop; thus

$$V_t = V_g - I_a R_a \tag{9.8}$$

However, when measuring the terminal voltage for a real machine, it is noted that the drop is greater than the IR drop. This additional drop is called the armature reaction. Its effect is shown in Figure 9.15.

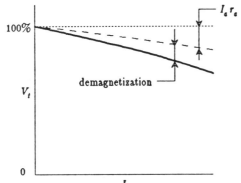

Figure 9.15 Separately excited generator load curve.

C. Shunt Generator Connection

A separately excited dc generator requires a variable dc power supply for the field. Since the energy required by the field is small compared with the load, some of the output may be diverted to the field circuit, as shown in Figure 9.14c. The machine is then said to be self-excited. However, there are several features that must be considered. These include the operating voltage of the shunt connection, the failure of the voltage build-up when the machine is started, and the possibility of reversed voltage polarity. Begin with the case of an unloaded shunt generator. A plot of the field resistance line is superimposed on a magnetization curve, as shown in Figure 9.16. It is observed in practice that the generator, if properly operated, will have a terminal voltage that corresponds to the intersection of the field resistance line with the magnetization curve. Thus, for a given shaft speed, the no-load terminal voltage may be varied by changing the resistance of the field circuit. This is generally accomplished by placing a variable resistance called a rheostat in the field circuit (see Figure 9.14c).

D. Voltage Build-up in a Self-excited Generator

A self-excited shunt generator will not necessarily provide a voltage output. It was this feature that kept the dc machine little more than a toy between the time of Faraday's machine in 1831 and Siemens' machine in 1866. The first machines had permanent magnetic fields; however, several inventors used a small permanent magnet generator to supply the field of a larger generator. Finally, Siemens showed that the machine could be self-excited

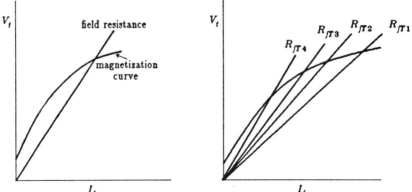

Figure 9.16 Shunt generator field resistance settings.

in 1866. The secret behind the self-excitation is the residual flux of the iron poles. That is, the magnetic flux remaining from the last time the machine was operated. Thus, when the machine is used for the first time or after a long storage period, there is a chance that the machine may not produce a voltage output. Also, the field connection must be proper, and the direction of rotation of the prime mover must be correct, or the machine will not build up an output voltage. Example 9.3, with the related Figures 9.17 and 9.18, lists the various possibilities for the three variables: (a) direction of the residual field, (b) shunt field connection, and (c) direction of rotation.

Example 9.3

Let us follow three of the eight possible cases. Begin with the condition in which the machine performs properly with the left pole having a south residual flux, a counterclockwise rotation, and field connection as shown in Figure 9.17a. Assume the residual flux produces a small terminal voltage with a positive potential at the plus terminal. With the field connection of Figure 9.17a, the small generator current produces an electromagnet that aids the residual flux. This increased flux produces a larger voltage. The larger voltage causes a larger field current with a related increase in field flux. This build-up continues to a point described by the intersection of the field line with the magnetization curve of Figure 9.16.

For a second case, consider the reversed residual (Figure 9.18 row b and Figure 9.17b). Initially, a voltage is induced in the armature with the polarity reversed from Figure 9.18, row a. This reversed polarity causes a current to flow in the field, which produces a flux to aid the residual.

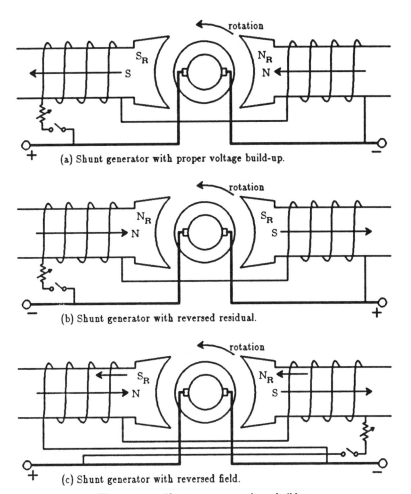

(a) Shunt generator with proper voltage build-up.

(b) Shunt generator with reversed residual.

(c) Shunt generator with reversed field.

Figure 9.17 Shunt generator voltage buildup.

Thus, the machine will build up to full voltage but with the polarity reversed from that of the original condition. If a machine is used in a factory where it is not desirable to reverse the terminals at every load in the plant, it will be necessary to change the residual flux, a process called "flashing the field." This may be done by connecting a battery to the machine terminals to establish the proper residual flux before starting.

A third case to consider is the reversed shunt field connection. Here, the polarity of the armature voltage caused by the residual flux is the same as for the original condition, a, of Figure 9.18. This voltage causes a

		Residual Field		Field Connection		Direction of Rotation		Voltage Build-up	
		SN	NS	Reg	Rev	CCW	CW	Pos	Neg
a	Original (Fig 9.17a)	X		X		X		Yes	
b	Reversed residual flux (Fig 9.17b)		X	X		X			Yes
c	Reversed field connection (Fig 9.17c)	X			X	X		No	
d	Reversed rotation	X		X			X		No
e	Reversed field and rotation	X			X		X		Yes
f	Reverse residual and field		X		X	X			No
g	Reverse residual and rotation		X	X			X	No	
h	Reverse residual, field and rotation		X		X		X	Yes	

Figure 9.18 Table showing voltage build-up for a shunt generator.

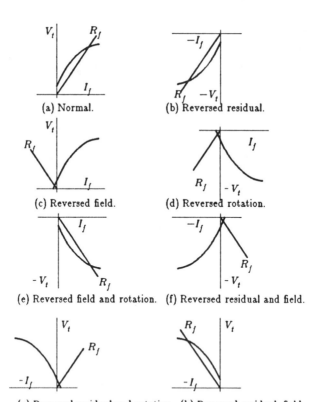

(a) Normal.

(b) Reversed residual.

(c) Reversed field.

(d) Reversed rotation.

(e) Reversed field and rotation.   (f) Reversed residual and field.

(g) Reversed residual and rotation.   (h) Reversed residual, field, and rotation.

Figure 9.19. Plot of magnetization curve and field resistance lines for the conditions of Figure 9.18.

current to flow in the field so as to produce a flux that opposes the residual flux, as shown in Figure 9.17c. This means that the terminal voltage of the generator shown in Figure 9.18c will be less than the terminal voltage without the field, which means the generator will not build up a voltage. Five additional conditions are included in Figures 9.18 and 9.19.

E. Field Discharge Circuit

The shunt field is a large electromagnet with a related large magnetic field energy. The energy that is stored in the magnetic field of a large machine can be destructive if discharged too rapidly. If a switch in the field circuit is simply opened, there is no discharge circuit and a high voltage arc will occur across the switch contacts. Eventually, the field insulation will be damaged and the machine will have to be replaced or repaired. Three methods used to provide a discharge circuit are:

1. A make-before-break switch to connect a resistor across the field winding. This type of switch prevents the resistor from draining off energy when the machine is in regular operation.

2. A nonlinear resistor (varister) across the field coil that offers high resistance at regular voltage and becomes a low resistance at higher voltage, thus draining off the magnetic field energy.

3. A diode circuit across the field winding that allows discharge but blocks current flow for regular operation.

F. Series Generator

A series generator that is operated at no load develops a small terminal voltage proportional to the residual flux. As the load current is increased, the flux increases and the terminal voltage increases. This continues until the field saturates and the generator becomes a constant current generator. It is true that for most dc series machines, this saturation occurs above the machine rating. However, the principle has been used successfully in supplying electrical energy for the old electric "street car" systems. It is also used for the "teeter-totter" mining systems in which the dc machine on the locomotive going down the hill will operate as a generator and the machine on the car going up the hill will operate as a motor.

G. Compound Generator

The compound generator includes the series and shunt fields. If the fluxes of the two coils aid each other, the machine is said to be *cumulatively compounded.* If the fluxes in the two coils oppose each other, the generator is said to be *differentially compounded.* Further, if there are enough turns of series windings so that the full-load voltage is greater than the no-load voltage, the generator is *over-cumulatively compounded* or simply *overcompounded.* If the series turns are of such a number that the no-load voltage is the same as the full load voltage, the generator is *flat-compounded.* If the turns are such that the full-load voltage is less than the no-load voltage, the generator is *under-compounded.* Most industrial plant dc generators are overcompounded. This means that the voltage will be almost constant throughout the plant. A flat compounded generator would provide a nearly constant potential at the generator terminals.

It is important to understand the characteristics of a differential generator. For example, I received a telephone call late one night from the electrician at the plant where I had been retained as a consultant. The plant was about 50 miles from my home. A small dc compound generator, which had been out for repair, had just been reinstalled. The electrician reported that the terminal voltage was proper at no load but that the terminal voltage dropped to almost nothing when the slightest load was connected. What would you have told the electrician to do? (See Problem 9.19.) Important applications for the differentially connected machine include dc arc welding generators and plasma-arc generators.

H. Summary of Generator Load Curves

The load curves for the various generator connections may be presented in several formats, as shown in Figure 9.20. There is a family of curves for each type of connection. For example, Figure 9.20a shows the load curves for a shunt generator for various values of the shunt field rheostat setting. This same kind of curve could be repeated for the other types of connections. This leads to the curves in Figure 9.20b, where the no-load potential is set the same for all except the series generator. Figure 9.20c is obtained by adjusting the full-load terminal voltages to be a common value for the various connections. Generally, the differential compound machine is shown on this type of figure.

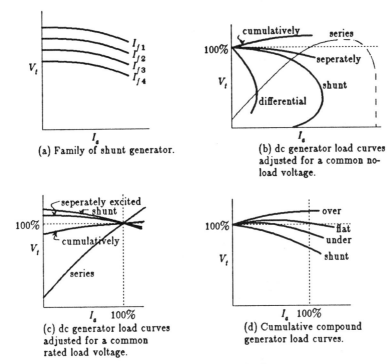

(a) Family of shunt generator.

(b) dc generator load curves adjusted for a common no-load voltage.

(c) dc generator load curves adjusted for a common rated load voltage.

(d) Cumulative compound generator load curves.

Figure 9.20 dc generator load curves.

9.7 dc MOTOR PERFORMANCE

A. Shunt Motors

The shunt motor load curves are generally plots of speed vs. armature current, as shown in Figure 9.22. The characteristics of a shunt motor with a shunt field rheostat are obtained by combining (9.8) with (9.2).

$$V_t = V_g + I_a R_a = K\phi\omega + I_a R_a \qquad (9.9)$$

Solve for

$$\omega = \frac{V_t - I_a R_a}{K\phi} \qquad (9.10)$$

A feature that is not necessarily obvious is described in (9.9). Note that as the flux is increased, the speed decreases. Also, as the flux is decreased, shunt field rheostat resistance is increased, and the motor speed increases. A family of curves for various values of shunt field resistance is shown in Figure 9.21. Note that the highest speed curve relates to the highest field resistance (lowest field current).

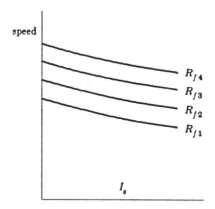

Figure 9.21 dc shunt motor load curves for various values of shunt field rheostat settings.

B. Series Motors

The load characteristics for a series motor are ideal for a load such as a diesel electric locomotive, where the required starting torque is large and the running torque relatively light, as shown in Figures 9.22 and 9.23.

C. Compound Motors

Figure 9.22 compares the speed curves for the various connections with the shunt and compound machines adjusted so that the no-load speed is a common value. When the speed curve increases with an increase in armature current, the machine is unstable. Generally, the differential machine is unstable for all values of armature current, as shown in Figure 9.22a. In Figure 9.22b, the characteristics are such that the differential and shunt are both unstable. Finally, there is a third possibility that a machine will provide a more constant speed if connected differentially as shown in Figure 9.22c. The relationship between torque and horsepower is given in Figure 9.23.

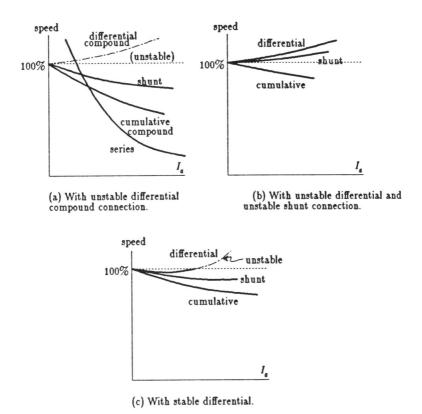

(a) With unstable differential compound connection.

(b) With unstable differential and unstable shunt connection.

(c) With stable differential.

Figure 9.22 dc motor load characteristics.

D. dc Permanent Magnet Motors

The speed-torque curve has a relatively steep slope as shown in Figure 9.24. For more information refer to the Bodine Motorgram [O.1, vol. 60, No. 5]

E. dc Motor Speed Control

One of the major advantages of the dc machine is the ease with which speed can be varied. An analysis of (9.10) reveals that the shunt machine may be varied by

    1. changing the field flux, which is normally accomplished by adjusting the field rheostat,

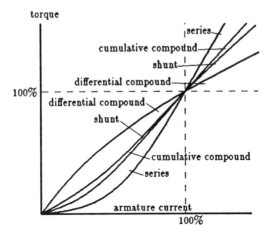

Figure 9.23 Comparative torques developed by dc motors.

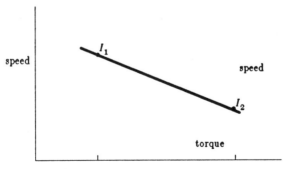

Figure 9.24 Permanent-magnet motor.

2. adding a series resistance in the armature circuit to change the total R_a, and

3. varying the terminal voltage, V_t.

Speed control of a shunt motor by adjusting the field rheostat gives a good speed control from approximately half-rated to maximum speed. On small machines it is often best to change the speed by varying the effective armature resistance by placing a rheostat in the armature circuit.

4. Another method of varying the shunt machine is the Ward-Leonard system shown in Figure 9.25. Here use is made of the principle of terminal voltage variation. This gives excellent speed control from

maximum speed through zero speed to reverse maximum. This system has the disadvantage of requiring the motor, a generator, and a drive for the generator. However, where speed control is required through an extreme range, this system may prove to be attractive. In recent times, controlled rectifier systems are replacing the motor-generator supply system. The controlled rectifier coupled to a separately excited dc shunt motor is called a *modified Ward-Leonard system*.

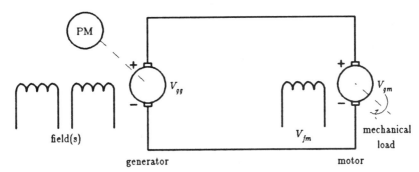

Figure 9.25 Ward-Leonard System.

F. Electrical to Mechanical Energy Transfer

The energy and power equations include the machine losses. The total input power is:

$$P_{in} = V_t I_L \tag{9.11}$$

The power delivered to the armature is:

$$P_a = V_t I_a = P_{in} - V_f I_f \tag{9.12}$$

The gross mechanical power is:

$$P_m = P_a - I_a^2 r_a \tag{9.13}$$

The shaft power is the gross mechanical power less the mechanical losses, as shown below:

$$P_{shaft} = P_m - \text{mechanical losses} \tag{9.14}$$

where the mechanical losses = friction, windage, and stray losses

G. Motor Starting Currents and Starting Resistances

When a dc motor is first connected to an electrical energy source, the current is limited by the armature circuit resistance as noted by (9.8) where

$$V_t = V_g + I_a R_a \tag{9.15}$$

But from (9.2), the generated voltage (or *back-emf*) is zero since the angular velocity is zero where

$$E_g = K\phi\omega = 0 \text{ at starting}$$

Thus,

$$I_a = \frac{V_t}{R_a}$$

For the machine of Example 9.1, this would be $I_a = 250/0.1 = 2500$ A. This current is 125 times the rated machine current. The heating and mechanical shock resulting from the related torque, where $\tau = K\phi i_a$ can be damaging to the machine, particularly if repeated each time the machine is started. It should be obvious that an additional resistance in the armature circuit during the starting operation would eliminate this problem.

For small fractional horsepower machines, the armature resistance is relatively high and the starting resistance is not always required. For the machines in the low horsepower range, a manual starter similar to that shown in Figure 9.26 is often used. In this device, the operator moves the handle slowly from the START position to the RUN position. When the wiper attached to the handle makes contact with the first wiper position, the shunt field is energized and resistance is placed in series with the armature. As the wiper is moved, there is less resistance in series with the armature, and the machine will accelerate. Then the handle is moved to the fully ON position, and a small electromagnet holds the handle in the RUN position. Note that when the wiper of the starter unit is in the START position, the shunt field is connected directly to the terminal voltage. As the wiper is moved toward the RUN position, the starting resistance is placed in series with the shunt field. Since the starting resistance is small, there is a small percentage of difference in field current between the two extremes. However, if the field winding were connected after the starting resistor while the armature current is flowing through, the starting winding would cause an appreciable difference in field current between the START and RUN position of the wiper arms. If terminal voltage is lost, the electromagnet will release, and the handle will be returned by a restraining spring to the OFF position. A note of caution: The starting resistance is rated for intermittent duty and should not be used for speed control purposes.

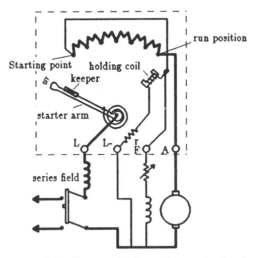

Figure 9.26 Wiring diagram of a four-point starting box for a dc compound motor.

For an automatic system that is started by depressing a pushbutton, the starting resistance is decreased in 2 to 5 discrete steps. The method for determining the values for the starting resistance is shown in Example 9.4.

There are many features of dc motors and generators that have not been presented in this introductory presentation. One related topic is presented in a later chapter. This is the use of a dc generator as a power amplifier mentioned in Chapter 12.

Example 9.4

Given the dc machine of Example 9.1, design the starting resistors for an automatic starting circuit where the starting current is limited to 225 percent of rated load current and is kept above the rated current.

Solution

Rated armature current is

$$I_{a\ rated} = \frac{50,000}{250} = 200 \text{ A}$$

Then 225 percent of rated current is:

$$2.25\,I_{a\,rated} = 450\text{ A} = I_{a\,max}$$

Beginning of Step 1. The steady state starting current is determined from (9.9) noting that the generated back voltage, V_g, of (9.2) is zero when the speed is zero.

$$V_t = V_g + I_{aH}r_{a1}$$

$$250 = 0 + I_{aH}r_{a1}$$

$$r_{a1} = \frac{250}{450} = 0.56\ \Omega$$

End of Step 1.

$$V_t = V_{g1} + I_{aL}r_{a1}$$

$$250 = V_{g1} + 200 \times 0.56$$

$$V_{g1} = 250 - 112 = 138\ \text{V}$$

Table 9.1				
Step	at beginning		at end	
	R_a (Ω)	V_g	V_g	ΔR (Ω)
1	0.56	0	138	0.31
2	0.25	138	200	0.14
3	0.11	200	228	0.01
4	0.1	228	---	----
	armature circuit resistance is 0.1 Ω			

Beginning of Step 2. The total resistance required at the beginning of Step 2 is calculated from (9.9).

$$250 = 138 + 450 \times r_{a2}$$

$$r_{a2} = 0.25\ \Omega$$

The resistance to be shorted out between Step 1 and Step 2 is (0.56 - 0.25) = 0.31 Ω, where 0.56 Ω is the resistance required at the beginning of Step 1 and 0.25 Ω is the resistance at the beginning of Step 2.

Repeat the above process to determine the resistance values for each step. The results are summarized in the Table 9.1. Note that on the fourth step all the external resistance is shorted leaving only the armature circuit resistance of 0.1 Ω.

H. Direction of Rotation

The direction of rotation for a dc motor is readily changed by reversing the current in the armature circuit with the relation to the current in the field circuit.

PROBLEMS

9.1 Name the major parts of a dc machine.

9.2 What are the major mechanical and electrical differences between a dc motor and a dc generator?

9.3 Name the major categories of dc generator connections. Sketch the circuit diagram for each of these connections.

9.4 Sketch the approximate curves of the terminal voltage (V_t) vs. armature current (I_a) for the various types of dc generator connections.

9.5 Plot a magnetization curve for the machine of Example 9.1 for a speed of 1500 rev/min.

9.6 If the shunt field current for the machine of Problem 9.5 is 5.5 A, how fast must the generator be driven to generate a terminal voltage of 240 V at no-load?

9.7 What is the required field current for the machine of Problem 9.5 when used as a generator at a no-load terminal voltage of 200 V at 1500 rev/min?

9.8 What size shunt field rheostat (resistance and current rating) is required for a shunt connection of the machine in Example 9.1 when it is driven by a constant speed synchronous motor at 1750 rev/min?

9.9 The magnetization curve of a dc generator contains the following points, all taken at a speed of 1000 rev/min:

Field current (A)	1.50	1.25	1.00	0.05
Induced voltage (V)	250	230	200	100

(a) If the field current is adjusted to 1.25 A, how fast must the generator be driven to generate 250 V at no-load? (b) What is the required field current for a no-load terminal voltage of 200 V at 800 rev/min? (c) If the machine is connected as a motor to a 230 V line and the field current is adjusted to 1.0 A, how fast will it run at no-load? Neglect rotational losses.

9.10 Explain step (d) of Figure 9.18.

9.11 Explain step (e) of Figure 9.18.

9.12 Five 250 V shunt generators have recently been returned from the repair shop. The first time these generators are driven by a common prime mover at rated speed they fail to build up to rated voltage. The indications of the five voltmeters connected to the armature terminals of the generators are listed in the accompanying table; the voltmeter indications are listed for when the field switches are open and closed. What should be done to have each generator build up to rated voltage with the correct polarity? (Correct polarity results when the voltmeter indicates up scale.)

Generator	Terminal Voltage (field switch open)	Terminal Voltage (field switch closed)
1	down scale	closer to zero
2	down scale	further down scale
3	+6	+3
4	+6	+20
5	+6	+6

9.13 Sketch the curves of speed vs. armature current for the various types of dc motor connections.

9.14 What is the theoretical maximum available torque of a series dc motor?

9.15 How do you reverse the direction of rotation of a dc shunt or dc series motor?

9.16 What is the maximum theoretical speed for a 1750 rev/min, 250 V, 20 hp, dc shunt motor?

9.17 Many students state that the direction of rotation of a dc shunt machine may be reversed by reversing the terminal corrections. Is this a true statement? Explain.

9.18 What are the major advantages and disadvantages of the dc motor: (a) shunt, (b) series, and (c) compound.

9.19 An electrician has installed a 125 V, 2 kW, compound generator. With no load, the terminal voltage is 121 V. When a load is connected to the generator, the terminal voltage drops almost to zero. What would you recommend the electrician do?

9.20 One method of changing the speed of a dc shunt motor is to vary the shunt field current by changing the setting of a shunt field rheostat. Is the resistance increased or decreased for an increase in motor speed?

9.21 Why must a starting resistor be used with all but the very smallest of dc motors? Why is the starting resistor not required on the small dc motors?

9.22 What are the three major types of dc motor speed control? What are the advantages and disadvantages of each?

9.23 What is the Ward-Leonard system? Describe any other equipment that could be used in place of the motor-generator set to supply the variable dc voltage source.

9.24 What is dynamic braking as applied to a dc motor?

9.25 Figure 9.27 shows the dc field circuit for a particular motor with a protective resistor R1 included. The purpose of the protective resistor (or field discharge resistor) is to keep the transient voltage across the field coil below a destructive value. The transient voltage of concern occurs when the switch is opened. What value of resistance is required for R1 in order to keep the voltage across the coil below 200 V when the switch is opened?

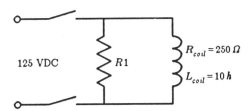

Figure 9.27 Shunt field circuit with protective resistor for Problem 9.25.

9.26 During the starting of a 25 hp, 250 V, dc shunt motor, it is desired that the armature current shall not exceed 2.5 times the full-load value of 85.0 A and that another step of the starting resistor be cut out whenever the armature current drops to its full-load value. The total armature circuit resistance is 0.16 Ω. Armature inductance is to be neglected. (a) How much external resistance must be inserted in series with the armature initially? (b) How much of this resistance may be cut out after the armature current drops to its rated value? (c) How much of the remaining resistance may be cut out in the second step? (d) How many total steps are necessary, and what are the values of resistance cut out for each step?

10

SINGLE-PHASE MOTORS

The majority of the small motors in use operate from a single-phase ac source. In general, these are relatively inexpensive devices providing rotary motion for home, commercial, and industrial use. These single-phase machines may be separated into five categories:

Single-phase, squirrel-cage induction

Universal (also called ac series)

Single-phase synchronous

dc permanent magnet with rectifiers (not described in this text)

Repulsion (not described in this text)

10.1 THEORY

The induction motor has no torque when stopped. There is just one phase of *breathing flux,* which was discussed in Chapter 4 and Figure 4.18. It would appear that the single-phase induction motor of Figure 10.1 cannot rotate. However, as was mentioned in Chapter 7, when an operating three-phase induction machine loses power to one of the three leads, the motor continues to run (usually overheating the machine). Thus, an induction motor will operate single-phase once started. How can this be? Three common theories have been developed to explain the single-phase motor operation. These include (1) double rotating field, (2) cross-field, and (3) symmetrical components [A.9, A.14]. For this presentation, only the double rotating field theory will be discussed.

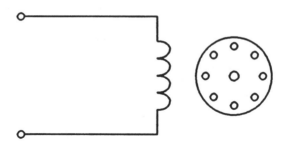

Figure 10.1 Circuit diagram for single-phase motor with main coil only.

Double Rotating Field Theory

Given a simplified one turn single-phase stator, the mmf in the air gap is:

$$F_1(\theta, t) = F_{(1\ peak)} \cos \theta \qquad (10.1)$$

For a sinusoidal current, the peak value of the mmf is expressed as:

$$F_{(1\ peak)} = F_{(1\ max)} \cos \omega t \qquad (10.2)$$

Combining (10.2) with (10.1) gives

$$F_1(\theta, t) = F_{(1\ max)} \cos\theta \cos\omega t \qquad (10.3)$$

The trigonometric identity of the product of two cosine terms allows (10.3) to be expressed as two terms

$$F_1(\theta, t) = (F_{(1\ max)}/2)(\cos(\theta - \omega t) + \cos(\theta + \omega t)) \qquad (10.4)$$

The expression of (10.4) consists of two rotating fields, each rotating in opposite directions. These fields produce the forward and backward torque speed curves of Figure 10.2. The sum of the two speed curves is the resultant torque-speed curve for an induction motor operating with only one single-phase winding. Note that the torque for zero speed (100 percent slip) is zero, as originally postulated. An induction motor will continue to rotate when supplied by single-phase if it can be started to rotate. There are six common methods for starting a single-phase motor. These are:

1. External torque method (a method normally used for testing only)

2. Capacitor start (induction run)

3. Split-phase

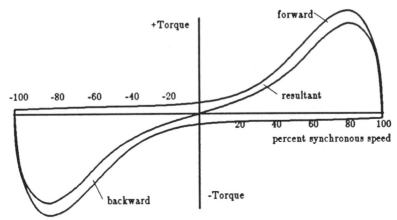

Figure 10.2 Torque-speed characteristic of a single-phase squirrel-cage induction motor with the main coil.

4. Permanent split capacitor-PSC (also called fixed capacitor)

5. Capacitor start-capacitor run (two capacitor)

6. Shaded-pole

10.2 PERFORMANCE CHARACTERISTICS OF SINGLE-PHASE INDUCTION MOTORS

A. External Torque Method

It should be obvious from the resultant torque-speed curve of the single winding induction motor, Figure 10.2, that if sufficient external torque is applied by a rope or strap to the motor shaft while rated voltage is connected to the main winding, the rotor will continue to revolve in the direction of the applied torque. This method provides a quick test when one of the other methods has a defective winding or switch.

B. Capacitor-start Motor

The stator of a capacitor-type induction motor (Figure 10.3) is wound as a two-phase motor with windings 90 electrical degrees apart. A capacitor is

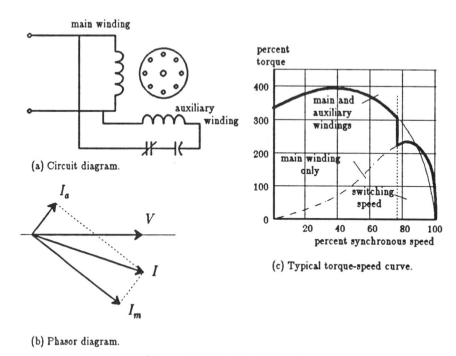

(a) Circuit diagram.

(b) Phasor diagram.

(c) Typical torque-speed curve.

Figure 10.3 Capacitor-start motor.

connected in series with one of the windings and the two windings connected to a single-phase power source. For the *capacitor start—induction run* (or simply *capacitor start*) type of motor, the capacitor is selected to produce a high starting torque. When the rotor reaches load speed, the capacitor does not provide the most desirable phase shift, so a centrifugal switch in series with the capacitor and the auxiliary winding opens so that the motor is allowed to operate in the single-phase mode. The change from capacitor run to induction run is accomplished by using a centrifugal switch that operates at about 75 percent of motor synchronous speed. The single-phase winding is called the *main winding* or *running winding,* while the coil with the capacitor is referred to as the *auxiliary winding, starting winding,* or *phase winding.* The capacitor adds expense to the motor and is required only for loads with higher starting torques.

The capacitor-start motor is used mainly where the load has relatively high starting torque such as for compressors and pumps. A comparison for different applications is given in Table 13.1.

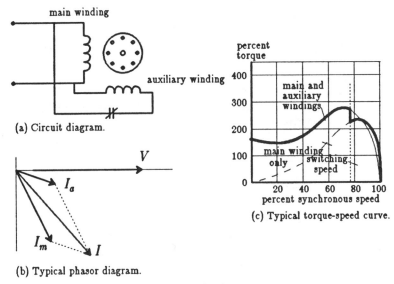

(a) Circuit diagram.

(b) Typical phasor diagram.

(c) Typical torque-speed curve.

Figure 10.4 Split-phase motor.

C. Split-phase Motor

If the auxiliary winding is wound with a smaller diameter conductor, the current in the two windings would be of a different phase shift and could be designed to produce adequate starting torque at less cost than using a capacitor (Figure 10.4). This technique provides what is referred to as the *split-phase motor*. Since the load torque produced by the main winding is better than that produced by the combined windings, a centrifugal switch opens the auxiliary winding at about 75 percent of synchronous speed. The auxiliary winding is rated for intermittent duty.

D. Permanent Split Capacitor (PSC) Motor

Applications requiring improved efficiency, which do not require a high starting torque, can use a correctly sized capacitor in the auxiliary winding (Figure 10.5). This winding is left permanently connected to the circuit and is referred to as the *fixed capacitor* or *permanent split capacitor* motor. The current in the auxiliary winding is 90 degrees out of phase with the main winding current. This means the motor is operating as a two-phase motor and the backward field is missing. Thus, the operation is quiet and

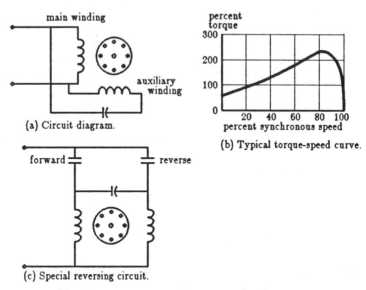

(a) Circuit diagram.

(b) Typical torque-speed curve.

(c) Special reversing circuit.

Figure 10.5 Permanent-split-capacitor (PSC) motor.

requires less current than other operations. This can be used for fans that operate for long periods of time.

E. Two-value Capacitor Motor

When the application requires the high starting torque of the capacitor start motor and the efficiency of the PSC motor, then the two capacitors connected as shown in Figure 10.6 provide an appropriate solution.

F. Shaded-pole Motor

For small motor applications, the shaded-pole design is worth consideration. A simplified version of this motor is shown in Figures 10.7 and 10.8. A slot is made in each pole of the stator. A coil or strap of electrical conductor called a *shading coil* is embedded around part of each pole face. The varying flux under the slotted section of the stator causes a current to flow in the shading winding. This current causes a countermagnetic flux so that the magnetic flux under the wrapped or shorted section lags behind the magnetic flux in the plain section. The combined magnetic fluxes create a

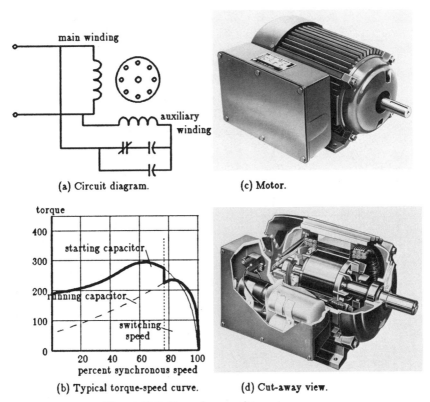

(a) Circuit diagram.

(c) Motor.

(b) Typical torque-speed curve.

(d) Cut-away view.

Figure 10.6 Two-value capacitor motor.

rotating magnetic field, small but adequate enough to cause the rotor to move. The shorting coil is referred to as a shade, hence the name *shaded-pole* motor. The shaded-pole motor produces a low starting torque; however, it can be built less expensively in small sizes. It is used in applications below 1/6 horsepower. Note that the split-phase motors are used for applications above 1/20 horsepower. Applications between 1/20 and 1/6 horsepower can use either the split-phase or shaded-pole, depending on cost and availability.

10.3 SINGLE-PHASE SYNCHRONOUS MOTORS

There are four common types of single-phase synchronous motors: reluctance, hysteresis, polarized synchronous, and synchronous permanent magnet. The last two are special applications of stepping motors and are discussed in Chapter 12.

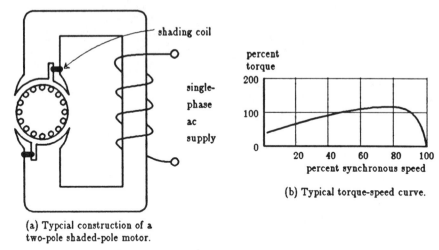

(a) Typcial construction of a
two-pole shaded-pole motor.

(b) Typical torque-speed curve.

Figure 10.7 Shaded-pole motor.

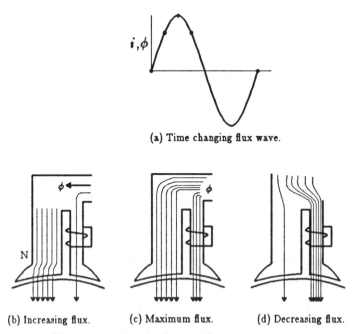

(a) Time changing flux wave.

(b) Increasing flux.   (c) Maximum flux.   (d) Decreasing flux.

Figure 10.8 Shaded-pole motor flux density patterns.

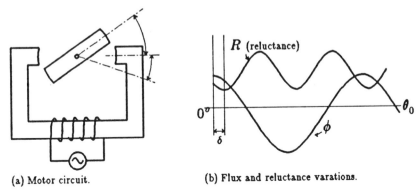

(a) Motor circuit.

(b) Flux and reluctance varations.

Figure 10.9 Elementary non-self-starting reluctance motor.

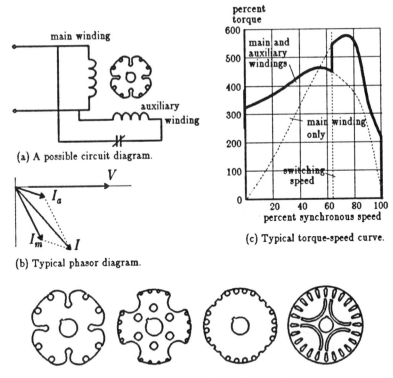

(a) A possible circuit diagram.

(b) Typical phasor diagram.

(c) Typical torque-speed curve.

(d) Typical rotor lamination punchings.

Figure 10.10 Self-starting reluctance-type synchronous motor.

A. Reluctance Motor

If the salient-shaped rotor of Figure 10.9 is standing still and an alternating current connected to the coil, the rotor will move in a position of minimum reluctance, as discussed in Chapter 2. This means that the rotor will line up with the axis of the poles with a pulsating force. It can be shown [A14B] that if the rotor is moved close enough to a speed in synchronism with the supply frequency, a torque of varying magnitude will be developed, but with an average unidirectional magnitude. The direction of rotation is determined by the direction of the starting motion. This non-self-starting method was used in many of the first electric clocks.

The reluctance motor can be made self-starting by providing any of the starting methods employed with the squirrel-cage induction rotor. The most common type of self-starting reluctance motor is the split-phase stator construction shown in Figure 10.10. A common type construction for a four-pole rotor is that shown in Figure 10.10. A disadvantage of the self-starting reluctance motor is that it is possible for a large load torque to prevent the rotor from synchronizing with the supply frequency.

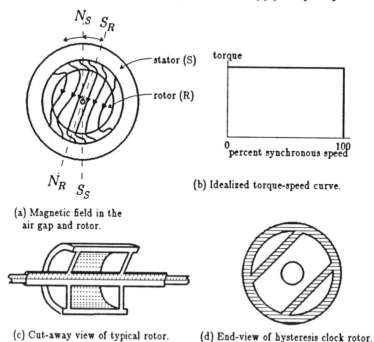

(a) Magnetic field in the air gap and rotor.

(b) Idealized torque-speed curve.

(c) Cut-away view of typical rotor.   (d) End-view of hysteresis clock rotor.

Figure 10.11 Hysteresis single-phase synchronous motor.

B. Hysteresis Motor

The rotor for a hysteresis motor is made with an outer ring of high-permeability steel, as shown in Figure 10.11. The stator must produce a constant rotating field. For single-phase machines this rotating field could be developed by a permanent split capacitor or by a shaded-pole. The rotating field causes a magnetic pole in the high permeability steel core, which by hysteresis develops the equivalent of a set of permanent magnet poles in the rotor. The rotor then follows the rotating force in the same manner that a polyphase synchronous motor does. One advantage of the hysteresis motor is that if the motor can develop enough torque to turn, it will turn at synchronous speed; otherwise, it will stall. A second advantage is that the torque is constant. The major disadvantage of the high-permeability rotor is the high initial cost.

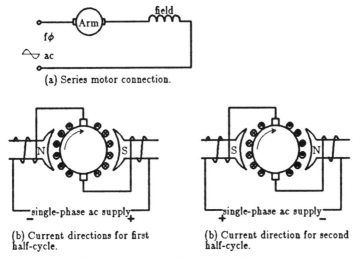

(a) Series motor connection.

(b) Current directions for first half-cycle.

(b) Current direction for second half-cycle.

Figure 10.12 ac series or universal motor.

10.4 UNIVERSAL MOTOR

The universal motor of Figure 10.12 is similar to a series dc motor except the stator and rotor magnetic circuits are constructed with laminated material so as to reduce the eddy currents. This means that the universal motor can be operated from either a dc or ac source. The three major advantages of the universal motor are (1) variable speed, (2) speeds higher than 3600 rev/min for the 60 Hz application, and (3) lower initial cost for high speed application. The universal motor requires a commutator and brushes; hence, increased maintenance.

10.5 DIRECTION OF ROTATION

The direction of rotation for a two-winding single-phase motor with squirrel-cage rotors is established by the original phase sequence. To change the phase sequence, interchange the connections to one of the pair of stator windings. Usually, the changes are made with the auxiliary winding. Note that the motor must be stopped and restarted for the windings to be effective in changing the direction of rotation for the split-phase, capacitor start, and the two-capacitor motors. The permanent split capacitor (PSC) can be reversed under load. The shaded-pole motor would require that the shade be placed on the other portion of the pole—an option not available with most standard designs.

10.6 SPEED VARIATION

The squirrel-cage rotor machine and the synchronous machines are not used for variable speed operation. However, the variable-pitch pulleys, gears, etc. mentioned in Chapter 7 can be employed to make them useful for variable speed operation.

The universal motor in conjunction with mechanical governors or SCR controllers make this type of machine an excellent choice for variable speed applications.

PROBLEMS

10.1 What is the shaft torque in newton-meters for a 1/6 hp, 1725 rev/min, split-phase induction motor?

10.2 Perform the following for a split-phase, single-phase induction motor: (a) sketch a circuit diagram, (b) sketch the speed-torque curve between 0 and 100 percent of synchronous speed, (c) describe how to change the direction of rotation, (d) describe how to change the speed.

10.3 Repeat Problem 10.2 for a capacitor-start, induction-run, single-phase motor.

10.4 Repeat Problem 10.2 for a PSC motor.

10.5 Explain how to reverse the direction of rotation for each of the following fractional-horsepower motors: (a) split-phase, (b) capacitor start-induction run, (c) fixed capacitor, (d) shaded-pole, (e) self-starting reluctance, (f) hysteresis, and (g) universal (or ac series).

10.6 Sketch the circuit diagram showing the necessary switching for reversing a 115 V, 1/3 hp, single-phase, split-phase induction motor.

10.7 What motor would you recommend for the following applications? Why? (a) A 3/4 hp drive motor for a water well pump with high starting torque. Power source: 240 V, single-phase, 60 Hz. (b) A 1/3 hp motor for turning the tumble tub for a natural gas clothes dryer. Power source: 120 V, 60 Hz, single-phase. (c) A strip-chart recorder for measuring pressure. Chart speeds at 5, 10, and 20 rev/min with a tolerance of +0.5 percent of specified speed. Power source: 120 V, 60 Hz, single-phase. (d) A drive motor for a small elapsed time counter.

10.8 What motor would you recommend for the following applications? Why? (a) The drive motor for a 1/20 hp residential kitchen exhaust fan. Power source: 120 V, 60 Hz, single-phase. (b) The 1/8 hp drive motor for a variable speed coil winding machine. Power source: 120 V, 60 Hz, single-phase. (c) A small 11,000 rev/min blower motor for an aircraft radar set. Power source: 120 V, 400 Hz.

10.9 What are the relative advantages and disadvantages of the different types of fractional horsepower motors that have squirrel-cage rotors?

10.10 What are the relative merits of the hysteresis motor as compared to the reluctance motor?

11

CONTROL OF
INDUCTION MOTORS

A small induction motor may be started with a manual switch with full line voltage applied to the motor terminals. The large machines are usually started with a magnetic starter that is operated by push-button control. In some applications, the starting currents are five to thirteen times the rated current. Under some conditions, the large line current may cause the terminal voltage to drop sufficiently so that other electrical equipment connected to the same source may be adversely affected. One exaggerated case was experienced by a small farming community of 1000 inhabitants who obtained their electric energy from a large utility system seven miles from the town. One farmer, whose feeding barn was near the town center, had a 20 hp motor that drew sufficient current to drop the voltage low enough to cause all the mercury arc street lights to turn off momentarily. If a mercury lamp is turned off, even momentarily, it takes approximately 5 minutes to regain full brilliance. Thus, the farmer was accused of turning off the street lights when he started his motor. The line voltage drop caused by a large current can cause excessive voltage dip at the motor terminals. This voltage dip can cause lights to dim, other motors to draw larger than normal currents, etc.

The motor starting circuit requires two basic types of protection: (1) protection against a line short circuit, and (2) protection against overload. Also, the circuit should have a means for disconnecting the motor for servicing. The next few paragraphs explain several starting circuits and describe the devices required for these circuits.

The short circuit protection for many small appliance motors is the branch circuit breaker located at the panel supplying the wall receptacle to which the appliance is connected. The plug on the appliance cord is a desirable and suitable circuit disconnect. In recent years, a thermal overload switch has been included in the frame of most small motors for longer motor life.

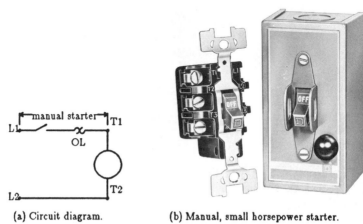

(a) Circuit diagram.   (b) Manual, small horsepower starter.

Figure 11.1 Manual starter circuit and devices for a small single-phase induction motor. *(Courtesy of the Square D Company)*

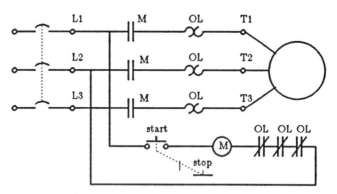

Figure 11.2 Circuit diagram of across-the-line magnetic starter and maintained-contact push button.

For slightly larger motors, 1/3 hp to 1 hp, manual starters that include thermal overload elements, Figure 11.1, may be used to switch the

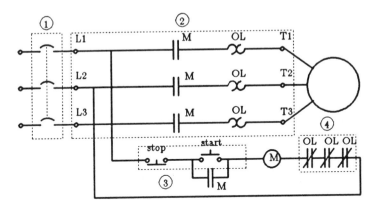

Figure 11.3 Circuit diagram of across-the-line starter and momentary-contact push button for providing low-voltage protection.

motor on and off. In cases where it is desirable to have remote control, then an electromagnetic contactor with the maintained contact switch circuit of Figure 11.2 is sometimes used. This circuit has the disadvantage that, in case of power failure, the motor may restart without warning when power is restored if the switch is left in the on position.

The automatic starting circuit of Figure 11.3 includes the basic elements of a simple full-voltage starting circuit. The circuit requires a circuit disconnect (1), contactor (2), push buttons (3), overload relays and contacts (4), etc. Consider the various devices used for operation of the circuit.

11.1 DEVICES USED FOR INDUCTION MOTOR STARTING CIRCUITS

An engineer should be familiar with the elements that form a motor control system. This system includes control functions as well as protective functions. The following is a listing of the components that are the "building-blocks" of the motor control system.

1. Line disconnect and branch short-circuit protection. This is usually a three-pole knife switch mounted in an enclosure (Figure 11.4) with appropriate fuses (Figure 3.19) or a three-pole circuit breaker in an enclosure (Figures 11.5 and 11.6).

2. Motor starter (contactor). This is an electromagnetically operated set of contacts that is used to establish or interrupt the electrical current to

(a) Open.   (b) Closed.

Figure 11.4 Disconnect switch. *(Courtesy of the Square D Company)*

the stator. Included with the contactor are thermal overloads and auxiliary contacts for special control functions (Figures 11.7 and 11.8).

3. Push buttons and switches. A typical push button of Figure 11.9a has two sets of contacts, one set normally open (NO) and the other set normally closed (NC). When the button is depressed, the circuit through the NC contacts is interrupted and the circuit through the NO contacts is completed. When the button is released, the switch contacts are spring returned to the normal position. For a stop-start station, two separate push buttons are required. Figure 11.9b shows several types of switches and push buttons that are available. These switch units may be assembled to provide multiple contacts. Assemblies of four and six contacts are quite common. The actuator unit may provide such features as: positive action 2, 3, and 4 position switches; spring return switches; push buttons with recessed buttons; push buttons with large mushroom buttons; keyed switches; etc.

4. Overload trip. The overload trip is a combination device that senses a protracted overload of the motors in the motor line circuit and opens the control circuit. The overload does not, indeed could not, open

Figure 11.5. Distribution circuit breaker panel with main breaker and eight feeder breakers. *(Courtesy of the Square D Company)*

the high current of the motor line circuit. These devices open the circuit to the contactor control and allow the high current contactor to open the motor lines. There are four types of thermal overload trips (sometimes called thermal overload relays or heaters).

The first is the *melting alloy thermal overload trip* of Figure 11.10 (also referred to as *solder pot relays*). In these overload trips, the motor current passes through a small heater winding. Under overload conditions the heat causes a special solder to melt, allowing a ratchet wheel to spin free, opening the contacts. Melting alloy thermal overload trips are *hand reset;* after they trip they must be reset by a deliberate hand operation. Thermal units are rated in amperes and are selected on the basis of motor full-load current, not horsepower rating.

The second type of thermal overload trip is the bimetallic thermal overload trip. Bimetallic thermal overload trips (Figure 11.11) employ a

Figure 11.6 Adjustable trip molded case circuit breaker, 100 ampere frame. *(Courtesy of the Square D Company)*

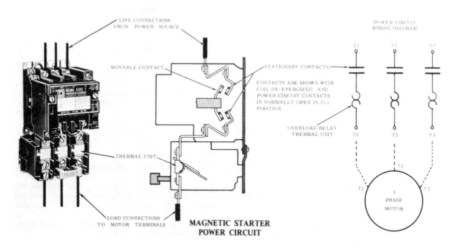

Figure 11.7 Line contactor or magnetic starter. *(Courtesy of the Square D Company)*

U-shaped bimetallic strip associated with a current-carrying heater element. When an overload occurs, the heat will cause the bimetallic strip to deflect and open a contact. These trips are available as hand or automatic reset. On automatic reset, the trip contacts, after tripping, will automatically

Figure 11.8 Cut-away views of motor starter control equipment. *(Courtesy of the Square D Company)*

reclose when the relay has cooled down. This is an advantage when the trip mechanisms are inaccessible. However, with this arrangement, when the overload trip contacts reclose after an overload, the motor will restart, and unless the cause of the overload has been removed, the overload relay will trip again. This cycle will repeat, and eventually the motor will burn out due to the accumulated heat from the repeated inrush current. More important is the possibility of danger to personnel. The unexpected restarting of a machine may find the operator or maintenance person in a hazardous situation as they attempt to find out why the machine has stopped.

The third type of overload trip is the magnetic overload relay (Figure 11.12). A magnetic overload relay has a movable magnetic core inside a coil that carries the motor current. The flux set up inside the coil pulls the core upward. When the core rises far enough (determined by the current and the position of the core), it trips a set of contacts on the top of the relay. The movement of the core is slowed by a piston working in an oil-filled dash pot (similar to an automotive shock absorber) mounted below the coil. This produces an inverse-time characteristic; i.e., the higher the current, the shorter the time delay before tripping. The effective tripping

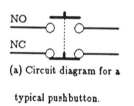

NO

NC

(a) Circuit diagram for a

typical pushbutton.

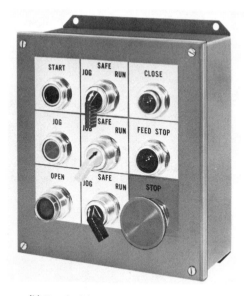

(b) Panel with several types of switches.

Figure 11.9 Push button and switch devices. *(Courtesy of the Square D Company)*

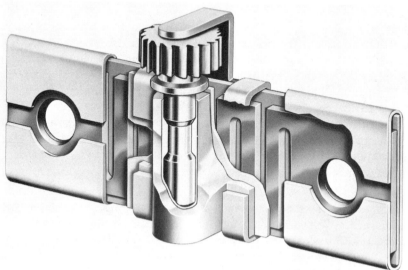

Figure 11.10 Melting alloy thermal overload. *(Courtesy of the Square D Company)*

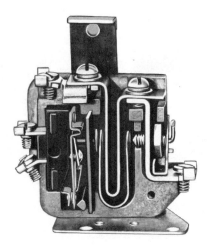

Figure 11.11 Bimetallic thermal overload. *(Courtesy of the Square D Company)*

current is adjusted by moving the core on a threaded rod. The trippir time is varied by uncovering the oil bypass holes in the piston. Because the time and current adjustments, the magnetic overload relay is som times used to protect motors having long accelerating times or unusu duty cycles. (The instantaneous trip magnetic overload relay is similar b has no oil-filled dash pot.)

The fourth type of overload trip is for motors over 1000 hp. F these motors it is necessary to use current transformers (Figure 11.13) sense the currents and then to use a system of sensitive protective relays.

5. Timers and timing relays (Figure 11.14). There are many typ and varieties of timing relays with many contact configurations. The relays may operate on the principle of (1) mechanical escapement, pneumatic-time delay, (3) hydraulic dash pot, (4) motor driven timers, current time delay in an RC circuit, or (6) electronic or solid-state circuit

11.2 FULL VOLTAGE STARTING

Now that the circuit devices have been described, return to the descript of the circuit of Figure 11.3.

Item 1. Short circuit protection and manual disconnect. This ma a fused disconnect (Figure 11.4) or a circuit breaker (Figure 11.6). T

Figure 11.12 Magnetic relay. *(Courtesy of the Square D Company)*

are four possible categories. The first is the instantaneous fuse (or non-time-delay fuse). This provides protection against a short circuit in the lines between the fuse and the motor terminals. The NEC® allows instantaneous fuse ratings from 150 to 300 percent of motor full-load current depending on the motor CODE letter, which is a nameplate designation of the starting current to rated current ratio. See NEC®[1], Tables 430-152 and 430-7(b). The overload protection is provided by the overload relays. The second type is the dual element fuse or time delay fuse described in Chapter 3. The dual element fuse has an instantaneous feature and a long-term overload feature. Since the fuse is rated in terms of the overload feature, the NEC® lists the overload limit for the dual element fuse at from 150 to 175 percent of motor full-load current. This provides some backup protection for the overload relays. The third type is the inverse time breaker. As was explained in Chapter 3, this breaker has a magnetic high current trip and a longer term thermal trip. The NEC® specifies a maximum rating of 150 to 250 percent of motor full-load current. The fourth

---

[1]National Electrical Code® and NEC® are registered trademarks of the National Fire Protection Association, Inc., Quincy, MA. Excerpts are included as Appendix C.

Figure 11.13. Current transformers for measuring large currents. *(Courtesy of the Square D Company)*

Figure 11.14 Time delay relay. *(Courtesy of the Square D Company)*

type is the instantaneous trip breaker. This is a breaker with only the magnetic high current trip, and the NEC® allows a maximum of 700 percent of motor full-load current. Do not jump to the conclusion that this is a poor choice. In reality, this is often the most safe and flexible method. Refer to NEC® section 430-152, which reads:

> An instantaneous trip circuit breaker shall be used only if adjustable, and if part of a combination controller having motor overload and also short-circuit and ground-fault protection in each conductor. A motor short-circuit protector shall be permitted in lieu of devices listed in Table 430-152 if the motor short-circuit protector is part of a combination controller having both motor overload protection and short-circuit and ground-fault protection in each conductor and if it will operate at not more than 1300 percent of full-load motor current. An instantaneous trip circuit breaker or motor short-circuit protector shall be used only as part of a combination motor controller which provides coordinated motor branch-circuit overload and short-circuit and ground-fault protection.

Item 2. Overload protection devices. This includes any of the four devices described in the paragraphs above. The heater or sensor is located in the motor line circuit and the trip contacts are located in the control circuit. The overload relays are to be selected to trip at 115 percent or 125 percent of the motor nameplate full-load current rating as given in NEC® section 430-32 (see Appendix C, paragraph 4). Also, three overload trips (one trip for each line) have been required on all new three-phase installations since 1971 as per NEC® Table 430-37.

Item 3. Contactor (Figures 11.7 and 11.8). The contactor is an electromagnet with electrical contacts attached to the magnet armature. The main contacts are designed to interrupt the motor line current and require appropriate *arc shoots* to quench the motor inductive current. Also, the contactor includes auxiliary contacts that are small and only rated for use in the control circuit. They serve control functions such as *seal-in* contacts.

Item 4. Control circuit. The power for the control circuit is normally obtained from one phase of the motor feeder. If the motor voltage is above 240 V, a step-down transformer is normally included. Two push button switches (Figures 11.3 and 11.9) are required, one for the START control and a second for the STOP control. When the START switch is depressed, the circuit to the contactor electromagnet coil is completed and the contactor will then close the motor feeder circuit. When the START button is released, the circuit would be returned to the non-operating mode except that a NO auxiliary contact of the contactor has closed around or seals in the starting circuit. This feature is a safety feature that provides loss of power protection. This means that on loss of power to the motor circuit, the contactor and the seal-in contact will open and make it necessary for the motor operator to manually restart the motor. If the START button were a maintained contact type, the motor would restart when power is restored to the lines and an operator could be injured when the motor is started unexpectedly. Depressing the STOP button will open the control circuit and stop the motor. Also, in case of overload, one or more of the overload trip units will open the control circuit and stop the motor. It is important to note that remote contacts that respond to pressure, motion, position, etc. could be provided in the control circuit to start and stop the motor.

11.3 RATING OF CONDUCTORS AND PROTECTION EQUIPMENT

The proper sizing of the fuses, circuit breakers, and conductors is not necessarily obvious. For example: a 100 A induction motor may be properly protected by a 700 A instantaneous trip breaker. This leads to a discussion of what are the limits and who sets the limits. The accepted standard for the United States is called the National Electrical Code® (NEC®).

Section 430 of the NEC® applies to motors, motor circuits, and controllers. A brief summary of the significant rules of the code follows; however, there are many special cases and problems that will not be considered. The important specifications relate to (1) circuit disconnect, (2) branch overcurrent protection, (3) motor circuit conductor, (4) motor controller and controller circuit, and (5) motor overcurrent protection.

All the ratings are based on the full-load current as given on motor nameplates, where available, or may be obtained from Tables 430-148, 430-149, 430-150, and 430-151 (see Appendix C).

The conductor size is normally rated at 125 percent of the motor full load amperes (NEC® Section 430-21) with the conductor sizes listed in NEC® Tables 310-16 and 18. Normally, type THW (thermoplastic, heat resistant, waterproof) insulation is used for conductors size #8 AWG and larger and type TW (thermoplastic, waterproof) insulation for conductors smaller than size #8 AWG.

Where the values for branch-circuit protective devices determined by Table 430-152 do not correspond to the standard sizes or ratings or to the possible settings of adjustable circuit breakers, the next higher size, rating, or setting is permitted.

Example 11.1

Given a 25 hp, Design B, Code Letter F, Service Factor 1.15, 460 V, 60 Hz, three-phase, squirrel-cage inductor motor, determine the (a) conductor size, (b) conduit size, (c) rating of the overload relays, and (d) ratings for the four types of short-circuit protection.

Solution

(a) The full-load current is found in NEC® Table 430-150 to be 34 A per phase. Thus, the conductor should be rated at 125 percent of 34 A or 42.5 A. Conductor size #8 AWG THW copper is required for the 42.5 A.

(b) From NEC® Table 3A of Chapter 9, a 3/4-inch conduit is required for the three #8 AWG conductors. An alternative is to use three #6 THW aluminum conductors in a 1-inch conduit. Note that only three power conductors are required for a motor since there is normally no neutral conductor. It is assumed that the conduit is the grounding conductor. In the case of a non-metallic conduit, such as PVC for underground use, a grounding conductor normally of smaller size (see NEC® Section 250-95) is required and the conduit size will have to be adjusted accordingly.

(c) According to NEC® Section 430-32, the motor overload relays should have a rating of 125 percent of the full-load current of 34 A or 43 A since the motor has a service factor of 1.15.

(d) The calculated limits for short-circuit protection are:

    (1) A non-time-delay fuse
        300% x 34 = 102 A. Use a 110 A fuse.
    (2) A dual-element fuse
        175% x 34 = 59.5 A. Use a 60 A fuse.
    (3) An instantaneous trip breaker in combination starter only,
        from (NEC® 430-152) 700% x 34 = 238 A.
    (4) An inverse time breaker
        250% x 34 = 85.0 A. Use a 90 A non-adjustable breaker.

Note that the answers for part (d) are maximum values. Many manufacturers recommend lower ratings in their catalogs.

11.4 REDUCED VOLTAGE STARTING

Machines connected at some distance from a supply transformer or those with large starting currents such as large Design A machines require reduced voltage starting equipment. The four most common types of reduced voltage starting schemes are:

1. Series resistance (Figure 11.15)

2. Series reactance

3. Autotransformer, often two transformers connected open-delta (Figure 11.16).

4. Part-winding connections

    a. Wye-delta starting

    b. Nine-lead delta

    c. Nine-lead wye

In considering the reduced-voltage systems, recall from (7.20) and (7.21) that the stator starting current is directly proportional to the applied voltage (11.1) and that the starting torque is proportional to the square of the applied voltage (11.2). Thus,

$$I_1 \propto V_1 \tag{11.1}$$

$$r \propto V_1^2 \tag{11.2}$$

Generally, Design B, C, and D motors and small Design A motors less than 7.5 horsepower can be started with full line voltage, often referred to as across-the-line or full-voltage starting. The large Design A motors and other special cases require reduced voltage starting.

The series type of starters are usually cheaper and do not cause a momentary loss of power during switching. However, the line current is equal to the motor current. For example, a series resistance which drops the voltage 50 percent will cause a reduction to 50 percent for the line current and 25 percent for the starting torque.

The autotransformer type starter causes a greater reduction in the line current because of the transformer turns ratio and is the preferred system for larger machines. For example, a turns ratio of 2 to 1 would reduce the motor voltage to 50 percent, with a resulting 50 percent motor starting current and 25 percent starting torque. However, the line current on the primary side of the autotransformer is 50 percent of the secondary current. Thus the line current is 50 percent of 50 percent or 25 percent of the normal starting current.

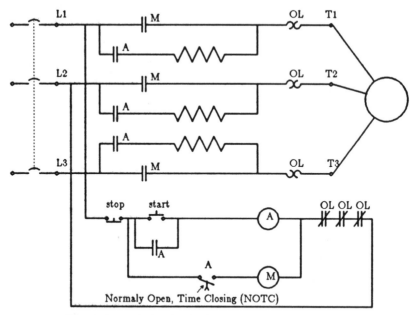

Figure 11.15 Reduced voltage starting using series resistance.

A form of reduced voltage starting can also be provided by part-winding motors, although the method does not involve reducing voltage at the motor terminals. Therefore, part-winding starting is not technically a reduced-voltage method but is generally included because of similarities in the control systems and in the advantages that are obtained with this method.

A part-winding motor has two separate, parallel windings. A part-winding starter consists of two ordinary starters, each selected for one-half the horsepower rating of the motor, and a time-delay relay. Pressing the start button connects one winding to the power supply. After the preset time delay, the second winding is also connected to the line. Full voltage is applied directly to each motor winding, but starting current and torque are substantially reduced by the motor design. This method of limiting starting current costs less than any of the reduced-voltage techniques. Part-winding starting is often favored when special power company regulations must be met. Although total available power may be large, momentary current limitations, usually specified in amperes for durations of 1 or 2 seconds, are imposed on the user to prevent excessive voltage disturbances. The main drawback of the part-winding starter is that torque may dip substantially during acceleration. Consequently, the motor may not fully

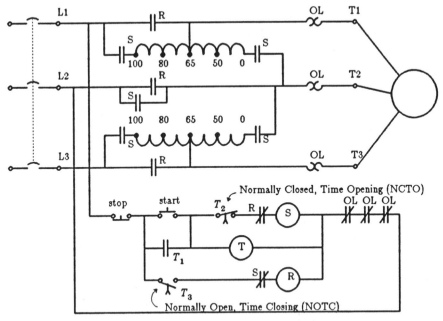

Figure 11.16 Reduced voltage starting using autotransformers.

accelerate a load on the first winding. Then, when the second winding is connected, it draws a large inrush current from the line and defeats the purpose of increment starting. Thus, part-winding starting is used to provide reduced increments of starting torque for particular machines or equipment and also to meet power company limitations. Typical applications include refrigeration systems, irrigation pumps, air conditioners, and conveyer systems.

11.5 ELECTRIC BRAKING AND PLUGGING

A standard across-the-line reversing starter can be used for quickly stopping an electric motor. With the motor running forward, the operator presses the reverse button. At the instant the motor reaches zero speed, the operator shuts off the power with the stop button. This action is called *plug stopping*.

Plug stopping is simple, inexpensive, and especially useful for such operations as lathe-turning; it doesn't matter if the motor makes a few reverse rotations before power is cut. However, if the nature of the machine or job prohibits the possibility of any reverse rotations, a zero-speed switch can be mounted on the motor shaft to assure power cut-off at zero speed. Sometimes current and torque developed during plugging may be too high, and series primary resistance, in the form of *plugging resistors,* is required to limit both to a proper value.

The main disadvantage of plug stopping is power line disturbance, especially by larger drives. When this presents a problem, another practical method that is frequently used for quickly stopping a squirrel-cage motor is dc dynamic braking.

11.6 REVERSING CONTROL

The rotating magnetic field of an induction motor determines the direction of rotor rotation. If the phase sequence is changed, the direction of rotation is changed. In practice this is usually accomplished by interchanging any two of the three stator leads. When it is desirable to accomplish the reversing by remote control, the circuit of Figure 11.17 may be considered.

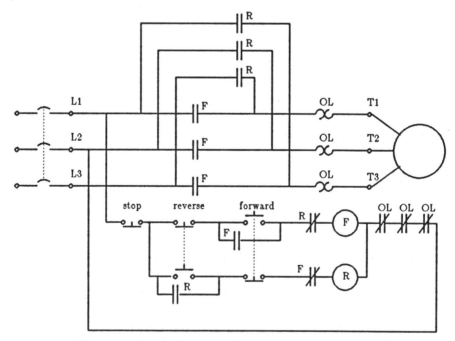

Figure 11.17 Reversing circuit for a squirrel-cage induction motor.

11.7 WOUND-ROTOR INDUCTION MOTOR

The wound-rotor machine is usually controlled by varying the resistance of the rotor. An automatic controller with a three-step acceleration is shown in Figure 11.18b.

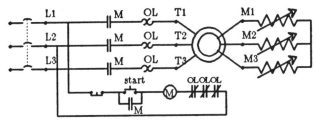

(a) Circuit for wound-rotor induction motor with a manually adjustable external rotor resistance.

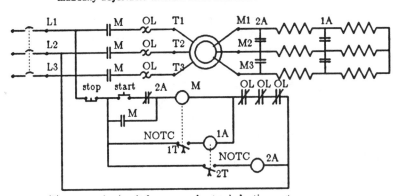

(b) Automatic circuit for a wound-rotor induction motor.

Figure 11.18 Control of a wound-rotor induction motor.

PROBLEMS

11.1 Draw the automatic starting circuit for a 100 hp, 460 V, 1725 rev/min, 60 Hz, three-phase, Design B, Code J, squirrel-cage induction motor that includes the following features: (1) one stop-start push button station, (2) across the line starting, (3) nonreversible.

11.2 Repeat Problem 11.1 but include the connection diagram for a second *stop-start* station 150 feet from the motor control contactor. (Note: If the remote control station requires more than three conductors, you have goofed.)

11.3 Draw the automatic starting circuit for a 100 hp, 460 V, 1110 rev/min, 60 Hz, three-phase, Design B, Code G, squirrel-cage induction motor that includes the following features: (1) autotransformer-type reduced voltage starting with 10 seconds time delay, (2) one stop-start station, (3) nonreversible.

11.4 Draw the automatic starting circuit for a 100 hp, 460 V, 1725 rev/min, 60 Hz, three-phase, Design B, Code J, squirrel-cage induction motor that includes the following features: (1) reversible, (2) across the line starting, (3) one "stop-start-reverse" station.

11.5 Given the motors of Table 7.2, specify the following for a 200 V application: (a) conductor size (copper with THW insulation), (b) conduit size, (c) size of dual element fuse, (d) size of inverse time circuit breaker, and (e) current rating of heaters (overload trip).

11.6 Given the motors of Table 7.2, specify the following for a 460 V application: (a) conductor size (copper with THW insulation), (b) conduit size, (c) size of dual element fuse, (d) size of inverse time circuit breaker, and (e) current rating of heaters (overload trip).

11.7 Specify the following for a system of three motors with motors A, D, and H of Table 7.2 operating at 200 V: (a) main feeder size to system (copper with THW insulation), (b) conduit for main feeder, (c) dual element fuse for each machine, (d) circuit breaker for main feeder, and (e) ampacity rating of heaters (overload trip) for each motor.

11.8 What is meant by reduced voltage starting?

11.9 What is the percentage of starting torque when an induction motor is started at 65 percent of rated voltage compared to the starting torque at rated voltage?

11.10 A three-phase induction motor has a starting torque of 10 N·m when 200 V is applied to the stator. What is the starting torque in N·m for a stator voltage of 150 V?

11.11 Given a 75 hp, 230 V, 60 Hz, three-phase induction motor with a full load current of 192 A. The starting current at full voltage is 5.5 times the full load current, and the starting torque is 500 N·m. Find the starting current and starting torque for a reduced voltage starting of 65 percent.

11.12 Given the motor of Problem 11.11, (a) determine the approximate magnitude of reduced voltage necessary to limit the starting current to three times rated current, and (b) determine the starting torque at this reduced voltage.

11.13 A three-phase squirrel-cage induction motor is started by means of a starting compensator (an open-delta autotransformer type of reduced voltage starter), which reduces the applied voltage to the motor to 50 percent of rated value. Determine (a) the starting torque as a percent of rated torque at rated voltage, (b) the stator starting current as a percent of the starting current at rated voltage, and (c) starting current in the supply mains leading to the motor compensator combination as a percent of starting current at rated voltage.

12

SPECIALTY AND
CONTROL DEVICES

Many special application and control machines have been developed over the years. A few new devices and some variations of old devices will be discussed in this chapter.

12.1 TWO-PHASE CONTROL MOTORS

Two-phase control motors are generally small induction motors designed to operate for long periods of time as positioning devices at or near zero speed. The rotor conductors have a relatively high resistance so that the maximum torque occurs at slips near 150 percent. A fixed or reference voltage is applied to the *main, fixed,* or *reference* winding. A controlling voltage that is displaced 90 electrical degrees from the reference winding voltage is

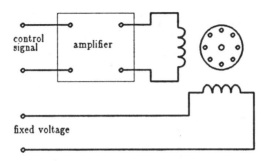

Figure 12.1 Schematic diagram of two-phase control motor.

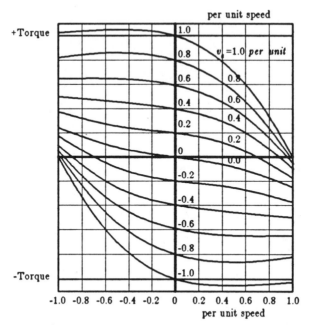

Figure 12.2 Typical torque-speed curves of two-phase control motor.

applied to the *control* or *signal* winding. By changing the magnitude of the voltage to the control winding, the shaft torque is varied. By changing the polarity of the control voltage, the direction of rotation may be adjusted. The torque-speed curves for various voltage values and phase relations is shown in Figure 12.2.

12.2 ac TACHOMETERS

If the shaft of an appropriately designed two-phase motor is connected to a rotating device and a voltage is applied to the reference winding, the voltage output of the control or signal winding is proportional to the speed of the rotating device.

12.3 STEPPER MOTORS

Stepping motors were first used by the British Navy to transmit shaft rotations in the early 1930's. Stepping motors were later used by the United

States Navy in instrumentation during World War II. Today stepping motors are used in a wide range of controlling systems: computer disc drives, printers, copy machines, electric typewriters, card sorters, carburetor adjusting, conveyors, silicon processing, milling machines, electron beam welders, lathes, and sewing machines. The stepper motor is a synchronous motor designed to operate at a low synchronous speed or from current pulses. It is excellent for discrete positioning steps and operates well with digital computer equipment. Three types of steppers will be mentioned here.

1. Mechanical steppers. Solenoid ratchet stepper motors are a type of mechanical stepper in which motion is caused by a solenoid action. The motor is locked (held in place) between pulses by a spring-loaded mechanical detent. The solenoid ratchet increments or decrements a step each time the spring is de-energized. Mechanical steppers provide a high torque for low stepping rates and are relatively inexpensive to manufacture. Their major disadvantage is that the mechanical parts wear and must be replaced.

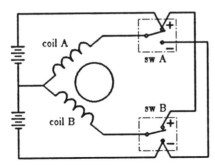

(a) Connections of the stator winding of a synchronous inductor motor.

Step	CW rotation		CCW rotation	
	Sw A	Sw B	Sw A	Sw B
1	+	+	+	+
2	-	+	+	-
3	-	-	-	-
4	+	-	-	+

(b) Combination of switching sequences.

Figure 12.3 Stepper motor test circuit and sequence.

2. Variable reluctance stepping motor (VR). The variable reluctance stepping motor does not have a permanent magnet in the rotor. It therefore relies on the magnetic field produced by the windings to create the force that moves the rotor. The variable reluctance motor understandably has very little residual torque when the power is removed. This stepper motor has a high torque to inertia ratio, high speed stepping rates, and longevity with bidirectional capability. There are two or more windings for the variable reluctance motors. The rotor teeth are magnetically independent of each other and are attached on the same rotor. The number of steps per revolution is determined by the number of teeth and windings in each phase.

N: number of steps per revolution = $T \times n$

T: number of teeth per phase

n: number of phases

R: number of degrees = $\dfrac{360^o}{Tn}$

When one of the phase windings is magnetized by a constant voltage, the electromagnetic force induces a current that moves the rotor teeth until they line up under the excited phase, which places the motor in equilibrium. Stepping motors generally operate on 200, 180, 144, 72, 24, or 12 steps per revolution, which generates an incremental shaft angle per step of 1.8, 2.0, 2.5, 5.0, 15, or 30 degrees. Stepping motors also do not have a zero or null position that can wear out a motor. Stepping motors may be rotated clockwise (CW) or counter-clockwise (CCW). They can be stopped and started at various mechanical rotational positions with the shaft of the rotor moving in precise angular increments. Stepping motors can operate in a closed or open loop system. A closed loop system uses feedback for correction or counting purposes, and an open loop system does not. Their highest torque occurs at the beginning of the step with a braking force at the end of the step. When the motors are idle they have a high holding torque that keeps the motor stationary between steps. They also have a low moment of inertia that allows for rapid starts, stops, and directional changes without a clutch or brake.

Stepping motors have the disadvantage of low efficiency. The loads must be carefully considered and stepping angles must be matched with the driven machine. For example, a disc drive out of position could cause damage to the disc or the drive.

3. Permanent magnet stepping motor (PM) or synchronous inductor motor. The motor was originally designed for driving loads at a slow

synchronous speed. The stator is constructed with a multiphase winding that can be energized by an external dc supply in various sequences to create a stepping motor action, as shown in Figure 12.3.

S: speed measured in rev/min $= \dfrac{60f}{N_r}$

f: electrical supply frequency measured in Hz

N_r: number of rotor teeth

12.4 SYNCHROS

Synchro is a generic name for a system of positioning devices that consist of a *transmitter* (or *generator*) and a *receiver* (or *motor)*. The synchro has been known by such names as *servomechanisms* and *self-synchronous drives*. Trade names for the synchro include *Autosyn* and *Selsyn*. The transmitter receives a rotary positioning input and converts it to electrical signals. The electrical signals are then transmitted by wire to the receiver, which converts the electrical signals to a rotary position. Applications for ships and large aircraft include remote compass indication, remote wind indication, rudder control, and aileron control. The synchro is also used for machine tool systems as position indicators. It is true that synchro systems are being replaced by analog-to-digital encoders, but there are still many synchro systems in operation.

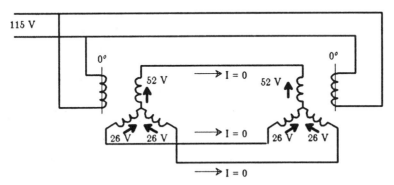

Figure 12.4 Synchro system voltages and currents when the synchros are 0° apart and the transmitter is set at 0°.

The single-phase synchro is a motor type device with a stator and a rotor. The transmitter and receiver have a stator with three sets of windings displaced by 120° and a salient pole rotor with two slip rings. These

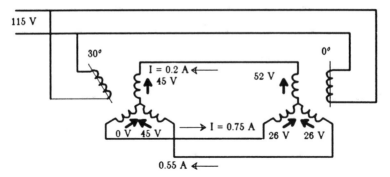

Figure 12.5 Synchro system voltages and currents when the synchros are 30° apart and the transmitter is set at 30°.

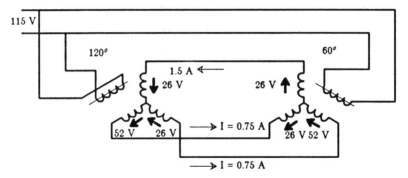

Figure 12.6. Synchro system voltages and currents when the synchros are 60° apart and the transmitter is set at 120°.

devices look much like a three-phase synchronous machine. The simple system consists of a transmitter and a receiver with the stators connected together as shown in Figure 12.4. A fixed frequency ac source is supplied to the rotors of the two units. For industrial and shipboard use this frequency is usually 60 Hz while 400 Hz is used in most aircraft applications. When the rotors of the two devices are at the same angular position, voltages induced in the stators are such that there is no current flow, as depicted in Figure 12.4. When the rotor positions are different with respect to each other, the stator voltages will be different, and a current will result causing a torque to each device as shown in Figures 12.5 and 12.6. If the transmitter is mechanically restrained, the receiver will move until the angle difference between the rotors returns to zero. There are three quirks to the system: (1) If two of the stator connections are reversed, the motion

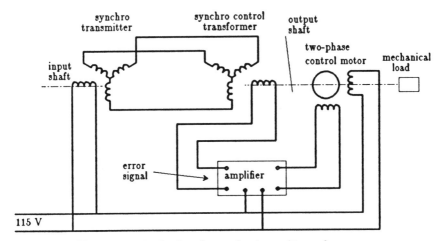

Figure 12.7 Application of a synchro transmitter and a synchro control transformer to angular position control.

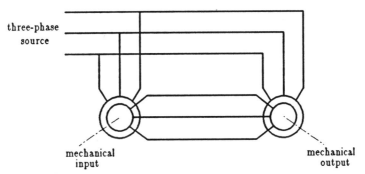

Figure 12.8 Two wound-rotor induction machines connected for use as a three-phase synchro system.

between receiving and sending will be opposite to each other; (2) at 180 degrees difference, there is a semistable zero torque; and (3) a mechanical damping is required on the receiver to prevent a step signal from causing the receiver to run continuously as a motor.

The size of the mechanical load that may be attached to the receiver is limited. This limitation can be greatly reduced by using a *synchro-control transformer* as the receiving device mechanically coupled to a two-phase control motor, as shown in Figure 12.7. The outside appearance of the control transformer is identical to that of the transmitter or receiver. The difference is that the rotor of the control transformer is made of many turns of fine wire to form a high impedance circuit. The output signal of the control transformer is then fed to an amplifier that supplies the control phase of the two-phase control motor. The control motor will then turn the mechanical load and the synchro control transformer shaft as long as the angular position of the rotors of the synchro devices are different.

Other synchro systems include differential units [G.7], which are used to add and subtract angles. Also, two three-phase wound-rotor induction motors connected as shown in Figure 12.8 will operate as a synchro positioning system.

12.5 RESOLVERS

The resolver is used for positioning feedback in many large machine tool systems. Many of these are being replaced by digital encoders, but for some systems the resolver is still considered more reliable. The resolver physically resembles a synchro, but internally it is different in that the rotor has two isolated windings wound at right angles to one another. The stator windings are also wound 90 degrees apart. A common application for the resolver is to apply a voltage C to one of the stator windings and to obtain a signal of $A = \cos \theta$ from one rotor winding and a signal of $B = \sin \theta$ from the other rotor winding. It is also possible to obtain a voltage of magnitude C from a stator winding with the rotor at a position of angle θ if known values of A and B have been applied to the rotor coils. Several other possibilities are explained by Ahrendt and Savant [G.1, p104].

12.6 LINEAR INDUCTION MOTORS

The linear induction motor is a machine that produces motion that is translational or linear, as opposed to the rotational motion found in the normally used electric motors. The principle of the linear motor was originally suggested in 1895 but was not sufficiently developed until the early 1960's. Currently linear induction motors are being used for high speed transportation and for material handling. One application of this motor was for a space equipment test vehicle that traveled at supersonic speeds. Small linear motors are used for sliding door closers, curtain pullers, and phonograph stylus arm drives.

The linear design may be explained by making a linearized view of the stator and rotor of a conventional polyphase induction motor. For this discussion, the member with the power windings is named the primary and the member with the conductor surface is named the secondary. It is possible to construct the system with either the primary or secondary as the stationary member. It should be obvious that the most practical systems use the secondary as the stationary member, which, just like trains, is limited by the length of the track.

For the linear case, the rotating magnetic field described in Chapter 4 moves linearly much like the waves in an ocean. This magnetic field induces currents in the secondary. The reaction field produces a force on the moving member and causes the appropriate motion.

There is a major drawback in the linear design. This is the force of attraction between the two fields. In the rotary design these forces are essentially balanced around the periphery. In the commercial linear machines, the magnetic forces have been as high as 10 times the forward electromagnetic force for the same area. In a simple one-sided machine, wheels are needed to support the vehicle and oppose the magnetic force. If the soft iron part of the secondary member is removed, the magnetic circuit is ruined for electromagnetic action as well as magnetic force, but if the soft iron part is separated from the conductive sheet, the magnetic pull is virtually eliminated. The next development was using the stationary soft iron structure as an active member. It was found that the soft iron structure could contain slots and be wound in a similar manner to the primary, producing a second primary member. This design enables the linear motor to carry twice as many conductors as its rotary counterpart, resulting in twice the output for a given primary member's heat loss. The two-sided linear motor operates with none of the friction and wind resistance of its rotary counterpart.

The synchronous velocity (velocity that matches the speed of the traveling wave) is expressed by

$$u_s = 2\tau f \qquad (12.1)$$

where τ is the pole pitch in meters (the distance measured between each pair of poles), f is the voltage source frequency in Hz, and u_s is the linear velocity in m/s. The velocity, u, of the moving member is less than synchronous velocity, u_s. This difference is defined as a ratio where

$$s = \frac{u_s - u}{u_s} \qquad (12.2)$$

Since the pole pitch τ can be made any desired value, the synchronous velocity for a given frequency can also have any desired value independent of the number of poles. The number of poles need not be even or even integral. Non-integral poles are those with more or less windings than the reference poles in the machine.

A second drawback is the increased air gap. The torque vs. slip curve of the rotary machine is replaced by a thrust vs. velocity curve for the linear machine. This curve is similar to that for a rotary machine with a high rotor resistance, i.e. maximum torque at high slip. The first reason for this high effective secondary resistance is the substantial area on the secondary sheet where orthogonally traversing currents do not exist. In these areas little thrust is produced. Much of this energy is dissipated as resistive losses without producing magnetic power. A second factor that produces high effective resistance is the air gap between the primary and secondary members. For the same power rating, this gap could be 25 times as large as that of a rotary induction motor. This larger gap requires much larger currents, and with the resistive losses discussed above, the machine operates at high slip and lower efficiency.

Another phenomenon unique to the linear induction motor which degrades its performance is the *end effect*. The linear motor has an *entry edge* at which new secondary conductors continually enter between the iron and the gap and an *exit edge* at which secondary conductors continuously leave. The motion of the secondary conductive sheet is a region of space in which the flux density varies so sharply (as it does at the entry and exit edges) that it causes the generation of eddy currents in the region of entry, which act to prevent the build up of the normal gap flux. Also, by Lenz's law, a current path tends to maintain the flux in the secondary after it has left the exit edge. This exit end effect causes only resistive losses that are insignificant compared to the entry end effect and therefore will not be considered. The influence of the end effect in high speed and low speed induction motors is quite different. The most important differences are as follows.

In the low speed motors, the speed of the end effect wave (the eddy currents) can be higher than the motor speed and even much higher than synchronous speed. In low speed motors, the attenuation (dissipation) of the entry end wave is quick, while in high speed motors the attenuation is very slow relative to motor speed. This slowly attenuating wave may be present over the entire length of the air gap. In the low speed motor, the end effect actually improves motor performance in the low slip region, which is the motor operating region. It increases thrust and efficiency and allows thrust to be developed even when motor speed is equal to or greater than synchronous speed. But in high speed motors, thrust and efficiency are greatly reduced in the low slip motor operating region. Because of the entry effect, high speed applications of linear induction may not be feasible unless methods to alleviate the end effect are used.

One method of alleviating the end effect is the use of compensating windings. The most economical type consists of a compensating winding with two poles placed in front of the entry end of the main winding. The exit end effect of the compensating winding is transmitted to the air gap of the main winding. The condition of compensation exists when the sum of the exit end wave of the compensating winding and the entry end effect wave of the main winding equal zero everywhere within length L_A. The compensating condition includes both amplitude condition and phase condition. The amplitude condition is met by controlling the number of turns in the compensating windings so that the amplitude of the compensating winding exit end wave is the same as the main winding entry end wave. The phase condition is met by controlling the delay zone length so that the phase of the compensating windings exit wave differs from the main windings entry end wave by 180 degrees, and the sum of the two waves is zero. If these conditions are met, the entry end effect is virtually eliminated, allowing for the use of linear induction motors in high speed applications.

12.7 ACYCLIC GENERATORS

An acyclic generator (also called a homopolar generator) is a dc generator with a rotor that cuts lines of magnetic flux in one direction only. Because the flux is cut in only one direction, the voltage output has no pulses (is acyclic) and is virtually ripple free. The acyclic generator can produce continuous load currents of 200,000 amperes at low voltages of 5 to 100 volts. They are capable of generating currents up to 500,000 amperes for one to seven seconds.

Michael Faraday produced the first acyclic generator, which consisted of a metal disc being rotated continuously in one direction between the poles of a permanent magnet. The dc voltage produced was picked up by sliding brushes.

Modern acyclic generators take the form of rotating discs or two concentric steel drums. The inner drum (or rotor) is machined from a solid forging. The outer drum is the stator. Magnetic flux is produced by two cylindrical coils concentric with the rotor. The flux enters the rotor through the main air gap at the center of the machine, passes axially underneath the collector, and leaves the rotor through the flux return gaps at both ends of the machine. The maximum voltage is produced between the two collectors that are toward the ends of the rotor. The current generated is conducted from the rotor into the stator through the collectors. The collectors have a rotating part on the rotor and a stationary part on the stator. Previously, the collectors consisted of a complex system of brushes. These brushes had high frictional losses that lowered the efficiency to as low as 50 percent. Now the current is collected by liquid metal between the stationary and rotating parts of the collectors. Although mercury has been used in some generators, an eutectic alloy of sodium and potassium, NaK, has proven to be the most effective. The eutectic alloy NaK is liquid at operating temperatures and has a lower viscosity than water. By using NaK, efficiency has been increased up to 98 percent. The NaK is prevented from oxidizing by pressurizing the generator with pure nitrogen.

The voltage produced by the acyclic generator is directly proportional to the field density and the speed at which flux lines are cut. The current produced by the acyclic generator is directly proportional to $D^{3/2}$ (D = rotor diameter). Current production capability is also limited by the generator's capability to dissipate heat.

Usually, a 10,000 kilowatt steam or gas turbine is used as a prime mover for each generator. Generators may be connected in series to increase the total voltage and current output. Acyclic generators only require 0.8 watts of excitation for every kilowatt of output.

The acyclic machine is also suitable as a motor with a very high power density (high horsepower per volume and weight).

12.8 AUTOMOTIVE ALTERNATORS

The alternator used to supply direct current for the automobile is a three-phase synchronous generator coupled with a three-phase bridge rectifier. The system is shown in Figure 12.9. The rotor is rather unique in that there is a single field coil (Figure 12.9b). Two soft iron assemblies are attached to each side of the coil. Note that one side will be magnetized as north poles and the other side as south poles. By the unique shape of the assemblies, there are alternating north and south poles to rotate inside the narrow three-phase stator assembly. This allows for a relatively inexpensive yet reliable system. The rectifiers (or diodes) are assembled in the end frame. The electronic regulator adjusts the current to the rotor (field) to maintain the appropriate terminal voltage.

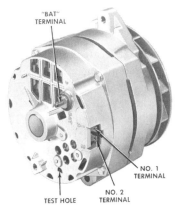

(a) Alternator assembly.

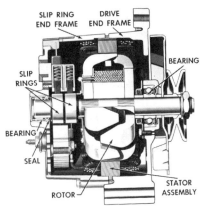

(b) Cross section view.

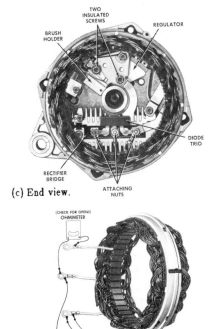

(c) End view.

(d) Stator winding.

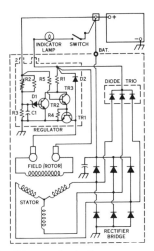

(e) Solid state circuit diagram.

Figure 12.9. Automobile alternator. (Courtesy Delco-Remy Division of the General Motors Corp.)

12.9 ROTATING AMPLIFIERS

Rotating amplifiers are created by cascading two dc generators with a feedback coil or coils for improved stability. These systems include machines marketed under the trademark of Rototrol and Regulex. By using a part of the armature winding to supply the field for the second stage, a system called a metadyne generator was developed. If the compensation is 100 percent, the device is known as an Amplidyne. A partial compensation machine was trademarked as an S generator. With no compensation, the machine was known as the Rosenberg generator, which was used to supply power for railroad passenger cars.

PROBLEMS

12.1 A vehicle is propelled by a linear induction motor. The vehicle contains a developed three-phase, eight-pole winding and is 4 meters long, 1 meter wide, and 1.7 meters high. The track on which the vehicle runs is a developed squirrel-cage winding. The 60 Hz power is fed to the car through contact arms extending through slots to rails below ground level. (a) What is the synchronous speed in kilometers per hour? (b) What is the synchronous speed in miles per hour? (c) What is the speed of the car for a slip of 25 percent in kilometers per hour?

12.2 Sketch the electrical circuit diagram for a wind direction indicator system that has a synchro torque transmitter (generator) connected to a wind vane and synchro torque receivers (motors) at two repeater stations.

12.3 Two dc generators electrically connected in cascade are driven by an induction motor. When driven at rated speed, dc generator #1 has an armature output of 1.0 A at 120 V when the separately excited shunt field is supplied with 20 mA at 60 V. When driven at rated speed, dc generator #2 has an output of 100 A at 120 V when the separatedly excited shunt field is supplied from generator #1 with 1.0 A and 120 V. What is the electric power gain of the system of the two cascaded generators, where the power of the field of generator #1 is the input, and the load connected to the terminals of generator #2 receives the output power?

13

APPLICATIONS

Motors and motor devices are selected or specified by mechanical, chemical, electronic, and computer engineers for equipment designed by these engineers. The motor type is determined by the load application such as starting torque, running torque, duty cycle, power, speed, rotary or linear motion, etc. The designer of the equipment using the motor is generally best suited to choose the correct motor.

13.1 MACHINE RATING

The nameplate rating of machines generally include seven major variables.

1. Rated kilowatts (or horsepower)
2. Temperature rise
3. Service factor (SF)
4. Duty cycle
5. Supply voltage and frequency
6. Speed (number of poles for an ac machine)
7. Torque characteristics

A. Rated Kilowatts (hp)

Rated kilowatts refers to the continuous output power rating guaranteed by the manufacturer. This value is limited by the heating of the motor parts caused by motor losses. These losses include stator and rotor winding losses, core losses including hysteresis loss and eddy current loss, bearing friction, windage or fan action of the moving rotor, and a small percentage of miscellaneous losses referred to as stray losses. The actual limit is the highest temperature rise of any part of the motor or generator. Obviously, the upper limit of power output of a machine operated in a desert climate will be lower than a machine operated in the arctic region. Likewise, the summer and winter capability of a given machine will be different. A high temperature can deteriorate the winding insulation and cause hot spots in the metal parts that can lead to fatigue, separation, or expansion of parts within the assembly. Motors have often been used on a continuous basis for 20 to 50 years.

B. Service Factor

Motors used to be given a service factor (SF) rating. This factor is the per unit rating at which the motor can be overrated for three hours without seriously decreasing the life of the motor insulation. For example, a manufacturer of large air conditioning equipment specified a 4160 V, 400 hp induction motor with a service factor of 1.25 to drive a 500 hp fan (400 x 1.25 = 500 hp). Two of these ac units were installed in a university student center. Insulation failure occurred on both machines eight years after installation with one machine failure occurring three weeks later than the other. The failure was a small, pin-sized hole in the insulation requiring a complete rewinding of the motor at a cost of $10,000 for each motor. At first it was thought the failure was due to lightning or ferroresonance. This kind of design error is one reason why the air conditioner manufacturer is no longer in business.

It might be interesting to note that electric utility companies often purchase large synchronous generators with a guaranteed rating, based on calculations, with a cost penalty if the machine does not measure up to standard. These large machines are loaded to the limit determined by temperature probes. Often the electric utility can get an extra 5 to 10 percent power from the machine without shortening the life of the machine.

C. Duty Cycle

There are three categories of motor duty cycle.

 1. Continuous duty

 2. Intermittent duty

3. Varying duty

Motors specified for continuous or intermittent duty are generally so marked on the nameplate. An example of an intermittent duty motor was one specified for use on a bomber anti-aircraft gun with 5 seconds on and 30 seconds off. It was calculated that 5 seconds was all the time available to fire at a high-speed target, and a second target would not be expected for another 30 seconds. The savings in weight for the smaller motor justified the design.

Different duty applications have loads that may vary between no load and some peak load. As the load varies, so does the machine temperature. "When there is a definite repeated load cycle, the motor size selection can be based on the rms value of motor losses for the load cycle. However, normally the losses at each increment of the load cycle are not available to the user. Therefore, a good approximation for the motor size selection can be based on the rms horsepower for the load cycle" [A.3, p100]. The following example illustrates the procedure.

Example 13.1

A motor is to be selected for a load that requires 40 hp for 12 minutes, 25 hp for 10 minutes, 15 hp for 15 minutes, and is stopped for 8 minutes. What size motor is required?

Solution

Calculate the square of the horsepower times the time for each of the time periods.

$$(hp)^2 \, t = [(40)^2 12] + [(25)^2 10] + [(15)^2 15] + 0$$
$$= 19{,}200 + 6{,}250 + 3{,}375 + 0 = 28{,}825 \text{ hp}^2{\cdot}\text{min}$$

The effective cooling time, taking into account 1/4 effective cooling when the motor is stopped, gives

$$t_{effective} = 12 + 10 + 15 + (1/4)(8) = 39 \text{ min}$$

Then

$$hp_{rms} = \sqrt{28{,}825/39} = 27.2 \text{ hp}$$

This means that a 30 hp motor based on heating should be considered rather than a 40 hp motor based on maximum load.

A serious error in motor application is oversizing (also called overmotoring). The disadvantages for the oversized design are

— lower efficiency,

— lower power factor,

— higher initial cost of the motor,

— higher initial cost of controls and feeders, and

— higher initial cost of installation.

Andreas [A3, p113] compares a 40 hp motor "that could have been selected based on the peak load versus the 30 hp motor that can be selected on the basis of the duty cycle:

1. Cost of motor: list price of standard 1800 rev/min drip-proof motor: 30 hp at $722; 40 hp at $908.

2. Cost of controls: NEMA-1, 240 V, general purpose motor starter: 30 hp, size 3 at $230; 40 hp, size 4 at $526.

This results in an initial capital investment difference of $482 (or 51%) excluding installation costs."

13.2 CONSTRUCTION

Enclosures. The standard housings for electric motors include open frame, drip proof (DP), totally enclosed fan cooled (TEFC), explosion proof, and submersible.

Bearings. Small motors often are furnished with sleeve bearings that provide a reliable general purpose system. Ball bearings have less friction but are more susceptible to damage if the motor shaft is bumped from the end. Roller bearings are used on larger motors. Very large motors use forced-oil bearings.

13.3 TYPES OF MOTOR LOADS

Fan. There are many types of fans. These include propeller, squirrel-cage, and impeller types. In general, the torque-speed curve for a fan is

$$\tau = k\omega^n \qquad \text{where n is generally between 2 and 4.}$$

Thus, when selecting a fan motor, the starting torque is small and the running torque is high. This means that a low starting torque induction motor is adequate for this application.

Compressors and pumps. These usually have high starting and running torques.

Conveyer systems. These have relatively high torque to initially move the system. Large rocks on ore conveyers can add additional difficulty at starting as they tend to jam the system.

Traction systems. These include diesel electric railroad locomotives, diesel electric mining trucks, golf carts, and electric cars. These require high starting torques, relatively low-running torques, and a readily adjustable speed. Good motors for this application include (a) dc series motors, (b) dc cumulative motors (in some applications), (c) Design D squirrel-cage motors, and (d) ac series motors.

Machine tools. These require all types of functions including constant speed, varying speed, constant torque, intermittent high torque, etc.

Cranes, hoists, elevators. These are considered vertical transportation and require motors similar to horizontal traction systems.

13.4 MOTOR ENERGY EFFICIENCY

As the relative cost of energy has increased over the past several decades, the need for energy conservation has increased. It has been estimated that between 60 and 70 percent of all the electric energy in the United States is used to supply electric motors. This means that the most effective electrical energy conservation would be accomplished by improving the efficiency of electric motors.

Large motors operate at above 90 percent efficiency, whereas many small motors are 60 percent or less. Two relatively modern methods for improving efficiency are the Nola (NASA) and Wanlass methods.

Two additional procedures for improving motor efficiency are (1) to use capacitors and (2) to consider energy efficiency designs of motor stators, rotors, etc.

Table 13.1 MOTOR APPLICATIONS

Characteristics given are typical — not limiting.

POLYPHASE MOTORS

Speed Regulation	Speed Control	Starting Torque	Breakdown Torque	Starting Current	hp Range	Power Factor	Percent Eff	Applications	Cost
SQUIRREL-CAGE (NEMA Design A)									
Low, 3-5%	Same as Design B.	Low, below 200 %	high, 200-250%	high, 500-1000%	0.5-200	High, 0.9	High 91-93	Limited application due to high starting current. Often requires reduced voltage starting.	Low
GENERAL-PURPOSE SQUIRREL-CAGE (NEMA Design B)									
About 3% for large to 5% for small sizes.	Adjustable frequency, multiple speed (2-4 constant speeds).	Good, 100% for 200 hp, 300% for 1 hp.	Good, 200% of full load.	Average, 500-550	0.5-500	Slightly lower than Design A.	91-93%	Most popular. For constant-speed service where starting torque is not excessive. Fans, blowers, rotary compressors, and centrifugal pumps, grinders, lathes, conveyers, process machinery, air-handling equipment. Usually full voltage starting.	Low
HIGH-TORQUE SQUIRREL-CAGE (NEMA Design C)									
About 3% for large to 7% for small sizes.	Same as Design B.	High, 250% of full load for high-speed to 200% for low speed.	200% of full load	Low, below Design B.	3-200	Less than Design A.	89-91%	Constant-speed where fairly high starting torque is required infrequently with starting current about 550% of full load. Reciprocating pumps, compressors, crushers, conveyers, etc.	Above Design B

Table 13.1 MOTOR APPLICATIONS (Continued)

Characteristics given are typical — not limiting.

POLYPHASE MOTORS (Continued)

HIGH-SLIP SQUIRREL-CAGE (NEMA Design D)

Speed Regulation	Speed Control	Starting Torque	Breakdown Torque	Starting Current	hp Range	Power Factor	Percent Eff	Applications	Cost
10-15%. Medium slip, (7-11%). High slip, (12-17%).	Same as Design B	High, 225 to 300% of full load.	200% at maximum slip.	Medium slip 400-800%. High slip 300-500%.	0.5-150	Low, 0.85	Low, 88-90%	Constant-speed and high-starting torque, if starting is not too frequent, and for high-peak loads with or without flywheels. Punch presses, shears, elevators, etc.	Above Design B

LARGE SQUIRREL-CAGE (Outside NEMA Design)

Speed Regulation	Speed Control	Starting Torque	Breakdown Torque	Starting Current	hp Range	Power Factor	Percent Eff	Applications	Cost
Custom designed with characteristics as required by application.					Up to 5000 hp and larger.			Constant-speed drives requiring economical and reliable operation.	Design related.

WOUND-ROTOR INDUCTION MOTOR

Speed Regulation	Speed Control	Starting Torque	Breakdown Torque	Starting Current	hp Range	Power Factor	Percent Eff	Applications	Cost
With rotor rings short circuited, 3-5%. With controller, 5-35%.	Speed can be reduced to 65% by rotor resistance.	Up to 300%	300% when rotor slip rings are shorted.	Depends upon external rotor resistances.	Typically 0.5-5000 hp, larger by special design.	High, same as Design A with rotor shorted, otherwise low.	Same as Design A with rotor shorted. Otherwise low.	Where high-starting torque with low starting current or where limited speed control is required. Fans, centrifugal and plunger pumps, compressors, conveyors, hoists, cranes, high inertia loads, etc.	High

Table 13.1 MOTOR APPLICATIONS (Continued)

Characteristics given are typical — not limiting.

POLYPHASE MOTORS (Continued)

POLYPHASE SYNCHRONOUS MOTOR

Speed Regulation	Speed Control	Starting Torque	Breakdown Torque	Starting Current	hp Range	Power Factor	Percent Eff	Applications	Cost
Constant	Adjustable frequency. Multiple speed (2-4 constant speeds).	40% for slow speed to 160% for medium speed 80% pf.	Unity pf motors 170%; 80% pf motors 225%; Specials up to 300%	Low	50-5000. Higher by special design.	Adjustable. May be rated for leading power factor.	High, 90 to 95%.	For constant-speed service, and where power-factor correction is desirable. Above 500 rev/min. Fans, blowers, dc generators, centrifugal pumps and compressors, reciprocating pumps, and compressors. Below 500 rev/min. Direct connection.	High for low hp. Low above 1000 hp.

SINGLE-PHASE MOTORS

SPLIT-PHASE -- General Purpose

Speed Regulation	Speed Control	Starting Torque	Breakdown Torque	Starting Current	hp Range	Power Factor	Percent Eff	Applications	Cost
10%	Adjustable frequency, multiple speed (2-4 constant speeds).	75% for larger sizes.	150% for larger sizes.	500-700%	1/20 - 2	Moderate	Moderate	Constant-speed service where starting is easy. Fans, saws, and grinders.	Low

SPLIT-PHASE -- High Torque

Speed Regulation	Speed Control	Starting Torque	Breakdown Torque	Starting Current	hp Range	Power Factor	Percent Eff	Applications	Cost
5-10%	Adjustable frequency, multiple speed (2-4 constant speeds).	Average, 200%	150%	Average, 500-700%	1/4 - 2	Moderate	Moderate	Washing machines, sump pumps, home workshops. For continuous and intermittent duty where operation is infrequent. May cause light flicker.	Moderate to low

Table 13.1 MOTOR APPLICATIONS (Continued)

Characteristics given are typical — not limiting.

SINGLE-PHASE MOTORS (Continued)

Speed Regulation	Speed Control	Starting Torque	Breakdown Torque	Starting Current	hp Range	Power Factor	Eff Percent	Applications	Cost
CAPACITOR START									
5% for large to 10% for small	Adjustable frequency, multiple speed (2-4 constant speeds).	150-350% of full load.	150%	Low	1/6 - 2	High	High	Constant-speed service for any starting duty and quiet operation where polyphase is not available. Ideal for all heavy duty drives such as compressors, pumps, and air conditioners. High starting torque with low starting current.	Moderate
PERMANENT SPLIT CAPACITOR (PSC)									
5%	Adjustable frequency, multiple speed (2-4 constant speeds).	Low, under 100%	150%	Low	1/6 - 3/4	High	High	For directly connected fan drives such as unit heater fans, air conditioner cooling coils fans, restaurant exhaust fans. Quiet operation. Not for belt drives.	Medium
TWO-CAPACITOR									
5%	Same as split-phase	High, over 300%	150%	Low	1/2-2	High	Medium	High starting torque of the capacitor start motor and the efficiency and quiet operation of the PSC.	High
SHADED POLE									
10% and greater	Poor	Low, under 100%	150%	High	1/300 - 1/6	Low	Low	Low power application. Small fans, blowers, inexpensive phonographs.	Lowest

Table 13.1 MOTOR APPLICATIONS (Continued)

Characteristics given are typical — not limiting.

SINGLE-PHASE MOTORS (Continued)

Speed Regulation	Speed Control	Starting Torque	Breakdown Torque	Starting Current	hp Range	Power Factor	Eff Percent	Applications	Cost
SERIES AC (UNIVERSAL)									
Variable speed.	Reduced voltage.	high, 500%	Variable	Variable with load torque	1/150-2	Varies	Varies	Sewing machines, portable tools, vacuum cleaners, and food mixers. Operates on ac or dc. High starting torque, high speed, varying speed regulation, small size and lightweight for given hp.	High for the same frame size. Low for higher speed with smaller frame size.

dc MOTORS

Speed Regulation	Speed Control	Starting Torque	Breakdown Torque	Starting Current	Applications	Cost
SERIES						
Varies inversely with load. Races on light loads and full voltage.	Zero to maximum.	High. Varies as square of voltage.	High.	Variable with load torque.	Where high-starting torque is required and speed can be regulated. Traction, bridges, hoists, gates, car dumpers, car retarders.	Low

Table 13.1 MOTOR APPLICATIONS (Continued)

Characteristics given are typical — not limiting.

Speed Regulation	Speed Control	Starting Torque	Breakdown Torque	Starting Current	Applications	Cost
SHUNT						
3 to 5%	Any desired range depending on design.	Good	High, 150-200%	Control regulated.	Where constant or adjustable speed is required and starting depending on conditions is not severe. Fans, blowers, centrifugal pumps, conveyers, wood and metal-working machines, and elevators.	Medium
COMPOUND (Cumulative)						
7 to 20%	Same as for shunt	Higher than for shunt, depending on amount of compounding	High, 160 - 200%	Control regulated.	Where high-starting torque and fairly constant speed are required. Plunger pumps, punch presses, shears, bending rolls, geared elevators, conveyers, hoists.	High
COMPOUND (Differential)						
Normally an unstable connection.						
SEPARATELY EXCITED						
Controllable to predetermined value.	Continuous full speed range in either direction.	Maximum of design.	200%.	Control regulated.	Precision speed and torque control with continuous full range control through reversed from full speed one direction to the other. Metal rolling mills, paper mills, ship propulsion, controlled processes.	High

PROBLEMS

13.1 Make a list of the different types of large horsepower motors. What are the relative advantages and disadvantages of each?

13.2 How do you change the speed of the following motors? (a) polyphase squirrel-cage induction motors (b) polyphase wound-rotor induction motors (c) polyphase synchronous motors (d) dc shunt motors (e) dc series motors.

13.3 How do you reverse the direction rotation of the motors listed in Problem 13.2?

13.4 What motor would you recommend for the following applications? Why? (a) A home storage wheat grinder which has a 1 hp motor. The unit runs on house current. (b) Bill Goodman, a friend who owns a service station, needs a replacement air compressor for his hydraulic hoist, tire air, and several pneumatic tools. He has found a good buy on a slightly used compressor, but it is lacking a 2 hp electric drive motor. The building is supplied with a 120/240 V, 60 Hz, four-wire, three-phase delta system. What motor would you suggest he consider? (c) The speed on the potter's wheel varies from 0 to 200 rev/min by means of a foot pedal attached to the wheel assembly by a cord. A 1/2 hp motor is used. Power is 120 V, single-phase, 60 Hz. Relatively constant speed at different speed settings is mandatory. What motor do you recommend? (d) What type of motors would you recommend for driving three separate movie projectors that are to be synchronized with each other to produce a three-section extra-wide-screen movie? (e) A chemical engineering research laboratory has need for a 5 hp variable-flow pump motor. The motor must operate at speeds between 100 rev/min and 3500 rev/min with a tolerance of 1.5 percent at any speed setting. The power supply available is three-phase, 60 Hz, at 480 V and 208 V.

13.5 [The following problem is similar to one used in an Engineer in Training (EIT) test.] A 20 hp, 1750 rev/min, three-phase, 460 V (line-to-line), 60 Hz induction motor is operating at full load output. Assume the motor efficiency is 90 percent and the power factor is 80 percent. Determine the following: (a) line current (b) kilovolt amperes of capacitor needed to increase the power factor to 90 percent, (c) approximate no-load speed.

13.6 The motor of Problem 13.5 is supplied from a 12.47 kV source through a 12,470:480 V, three-phase transformer. (a) What is the magnitude of the transformer primary feeder currents with and without the capacitor? (b) What is the transformer kVA rating with and without the capacitor?

13.7 What motor would you recommend for the following applications? Why? (a) A 100 hp air conditioning blower fan for a large building. Speed approximately 1100 rev/min. The power source is 2300 V, 60 Hz, three-phase. (b) What motor would you select for use on a 10 hp food canner machine where the speed must be varied between 2000 rev/min and 3500 rev/min? The power source is 240 V, three-phase, 60 Hz. The starting torque is 125 percent of rated torque. (c) A gravel sorter in a cement batch plant is a long slanted cylinder of wire mesh. Washed gravel is fed in at one end. As the tube rotates, the smaller gravel immediately falls through the mesh. Further down the inclined tube, the larger stones fall through a coarse mesh and so on. A 15 hp, three-phase, 480 V, 60 Hz motor is suggested. What type should be recommended? The motor must be able to start with stones in the cylinder.

13.8 What motor would you recommend for the following applications? Why? (a) Select a motor or motor system for use with a vibration test stand where the speed must be varied between 10 rev/min and 1000 rev/min. The maximum power requirement at the high speed is 25 hp. The speed at any setting must be held to plus or minus 10 percent of set speed. The power source is 240 V, 60 Hz, three-phase. The starting torque is 125 percent of rated full-load torque. (b) A 5 hp airconditioning compressor motor. The power source is 208 V, 3-phase, 60 Hz. (c) A 100 hp coal crusher is used to provide pulverized coal for the boiler of a steam plant. The crusher is normally started with the hoppers full of coal. The power supply is 480 V, three-phase, 60 Hz. What motor would you recommend?

13.9 Which motor would you recommend for the following applications? Why? (a) Tile and brick saw. The 2 hp, single-phase motor is operated at 120 V or 240 V, 60 Hz. It is used to cut brick, cement blocks, and tile and is often started on reduced voltage due to losses in the long extension cord. (b) A "pipe thread cutter" supplies a constant speed at about 20 rev/min. The motor must operate from 120 V, single-phase, 60 Hz. High torque is often required in starting and running. The motor is reversible.

13.10 What is the cost per month for operating the motor of Example 13.1? Consider two 8-hour shifts a day for 23 days per month with energy costs (a) at 6 cents per kWh and (b) at 10 cents per kWh. Consider the 30 hp motor efficiency to be 70 percent at 15 hp, 92 percent at 25 hp, and 80 percent at 40 hp.

13.11 Repeat Example 13.1 for the motor operating with the following loads and times:

60 hp for 2 min
30 hp for 10 min
5 hp for 20 min
stopped for 12 min

13.12 Write a problem similar to Problems 13.7-13.9. Submit this to the instructor on a 3 inch × 5 inch card. Include motor nameplate data for your example or describe your own recommendations.

APPENDIX A

SYMBOLS AND ABBREVIATIONS

Quantity	Name	Unit Abbreviation	Symbol	In Terms of other units
	alternating current	ac		
wire size	American Wire Gauge	AWG		
electric current	ampere	A	i, I	base
	ampere-turn	At	(NI)	
wire size	circular mil	cmil		
	thousand circular mil	kcmil (also MCM)		
	direct current	dc		
capacitance	farad	F	C	Q/V
inductance	henry	H	L	Wb/A
frequency	hertz (cycles per second)	Hz	f	s^{-1}
power	horsepower	hp		
energy, work	joule	J		N m, W s
mass	kilogram	kg	m	base
length	meter	m	l	base
force	Newton	N	F	m kg/s^2
energy, work	Newton-meter	N m		W s, J
torque	Newton-meter	N m	T	
resistance	ohm	Ω	R	A/V
	power factor	pf		
	radian	rad		
	radian per second	rad/s	ω	
	revolution per minute	rev/min	n	
	revolution per second	rev/s		
conductance	siemens (mhos)	S	G	A/V
time	seconds	s	t	base
magnetic flux density	tesla	T		Wb/m^2
	var	var	Q	
voltage	volt	V	v, V	W/A
apparent power	volt-amperes	VA	S	
power	watt	W	p, P	J/s or N m/s
energy(work)	watt-second	W s		J, N m
	watt-hour	Wh		
magnetic flux	weber	Wb	ϕ	

APPENDIX B

EXCERPTS FROM THE NATIONAL ELECTRICAL CODE

1. Conductor Size (NEC® 220-2(a))

Branch Circuits. "Continuous Loads. The continuous load supplied by a branch circuit shall not exceed 80 percent of the branch-circuit rating."

2. Protection of Conductors (NEC® 240-3)

"Conductors, other than flexible cords and fixture wires, shall be protected against overcurrent in accordance with their ampacities as specified in Tables 310-16 through 310-19 and all applicable notes to these tables."

3. Standard Fuse and Circuit Breaker Ratings (NEC® 240-6)

The standard ampere ratings for fuses and inverse time circuit breakers shall be considered 15, 20, 25, 30, 35, 40, 45, 50, 60, 70, 80, 90, 100, 110, 125, 150, 175, 200, 225, 250, 300, 350, 400, 450, 500, 600, 700, 800, 1000, 1200, 1600, 2000, 2500, 3000, 4000, 5000, and 6000.

Exception: Additional standard ratings for fuses shall be considered 1, 3, 6, 10, and 601.

4. Notes to Tables 310-16 through 310-19

The following is an excerpt of the notes for the related tables.

8. More Than Three Conductors in a Raceway or Cable. Where the number of conductors in a raceway or cable exceeds three, the ampacities given in Table 310-16 or 310-18 shall be reduced as shown in the following table:

Number of Conductors	Percent of Values in Tables 310-16 and 310-18 as Adjusted for Ambient Temperature if Necessary
4 thru 6	80
7 thru 24	70
25 thru 42	60
43 and above	50

Where single conductors or multiconductor cables are stacked or bundled longer than 24 inches (610 mm) without maintaining spacing and are not installed in raceways, the ampacity of each conductor shall be reduced as shown in the above table.

Exception No. 1: When conductors of different systems, as provided in Section 300-3, are installed in a common raceway the derating factors shown above shall apply to the number of power and lighting (Articles 210, 215, 220, and 230) conductors only.

Exception No. 2: The derating factors of Sections 210-22(c), 220-2(a), and 220-10(b) shall not apply when the above derating factors are also required.

Exception No. 3: For conductors installed in cable trays, the provisions of Section 318-10 shall apply.

Exception No. 4: Derating factors do not apply to conductors in nipples having a length not exceeding 24 inches (610 mm).

9. Overcurrent Protection. Where the standard ratings and settings of overcurrent devices do not correspond with the ratings and settings allowed for conductors, the next higher standard rating and setting shall be permitted.

Exception: As limited in Section 240-3.

Table 310-16

Ampacities of Insulated Conductors Rated 0-2000 Volts, 60° to 90°C
Not More Than Three Conductors in Raceway of Cable of Earth (Directly Buried),
Based on Ambient Temperature of 30°C (86°F)

Size	Temperature Rating of Conductor							
	60°C	75°C	85°C	90°C	60°C	75°C	85°C	90°C
Types	†RUW †T †TW †UF	†FEPW †RH †RHW †RUH †THW †THWN †XHHW †USE †ZW	V M	TA TBS SA AVB SIS †FEP †FEPB †RHH †THHN †XHHW§	†RUW †T †TW †UF	†RH †RHW †RUH †THW †THWN †XHHW †USE	V M	TA TBS SA AVB SIS †RHH †THHN †XHHW§
	Copper				Aluminum or Copper-clad Aluminum			
AWG								
18	....	....	....	14	....	....	....	....
16	....	....	18	18	....	....	....	....
14	20†	20†	25	25†	....	....	....	....
12	25†	25†	30	30†	20†	20†	25	25†
10	30	35†	40	40†	25	30†	30	35†
8	40	50	55	55	30	40	40	45
6	55	65	70	75	40	50	55	60
4	70	85	95	95	55	65	75	75
3	85	100	110	110	65	75	85	85
2	95	115	125	130	75	90	100	100
1	110	130	145	150	85	100	110	115
0	125	150	165	170	100	120	130	135
00	145	175	190	195	115	135	145	150
000	165	200	215	225	130	155	170	175
0000	195	230	250	260	150	180	195	205
MCM								
250	215	255	275	290	170	205	220	230
300	240	285	310	320	190	230	250	255
350	260	310	340	350	210	250	270	280
400	280	335	365	380	225	270	295	305
500	320	380	415	430	260	310	335	350
600	355	420	460	475	285	340	370	385
700	385	460	500	520	310	375	405	420
750	400	475	515	535	320	385	420	435
800	410	490	535	555	330	395	430	450
900	435	520	565	585	355	425	465	480
1000	454	545	590	615	375	445	485	500
1250	495	590	640	665	405	485	525	545
1500	520	625	680	705	435	520	565	585
1750	545	650	705	735	455	545	595	615
2000	560	665	725	750	470	560	610	630

†The overcurrent protection for conductor types marked with an obelisk (†) shall not exceed 15 amperes for 14 AWG, 20 amperes for 12 AWG, and 30 amperes for 10 AWG copper; or 15 amperes for 12 AWG and 25 amperes for 10 AWG aluminum and copper-clad aluminum after any correction factors for ambient temperature and number of conductors have been applied‡.

§For dry locations only. See 75°C column for wet locations.

‡See the complete table in the NEC® for temperature correction factors.

10. Neutral Conductor.

(a) A neutral conductor which carries only the unbalanced current from other conductors, as in the case of normally balanced circuits of three or more conductors, shall not be counted when applying the provisions of Note 8.

(b) In a 3-wire circuit consisting of 2-phase wires and the neutral of a 4-wire, 3-phase wye-connected system, a common conductor carries approximately the same current as the other conductors and shall be counted when applying the provision of Note 8.

(c) On a 4-wire, 3-phase wye circuit where the major portion of the load consists of electric-discharge lighting, data processing, or similar equipment, there are harmonic currents present in the neutral conductor and the neutral shall be considered to be a current-carrying conductor.

11. Grounding Conductor. A grounding conductor shall not be counted when applying the provisions of Note 8.

NEC® Chapter 9. Table 3A Maximum Number of Conductors in Trade Sizes of Conduit or Tubing (Based on Table 1, Chapter 9)													
Conduit Trade Size (inches)		½	¾	1	1¼	1½	2	2½	3	3½	4	5	6
Type Letters	Cond Size												
TW, T, RUH, RUW, XHHW (14 thru 8)	AWG												
	14	9	15	25	44	60	99	142					
	12	7	12	19	35	47	78	111	171				
	10	5	9	15	26	36	60	85	131	176			
	8	2	4	7	12	17	28	40	62	84	108		
RHW and RHH (without outer covering), THW	14	6	10	16	29	40	65	93	143	192			
	12	4	8	13	24	32	53	76	117	157			
	10	4	6	11	19	26	43	61	95	127	163		
	8	1	3	5	10	13	22	32	49	66	85	133	
TW, T, THW, RUH (6 thru 2), RUW (6 thru 2),	6	1	2	4	7	10	16	23	36	48	62	97	141
	4	1	1	3	5	7	12	17	27	36	47	73	106
	3	1	1	2	4	6	10	15	23	31	40	63	91
	2	1	1	2	4	5	9	13	20	27	34	54	78
	1		1	1	3	4	6	9	14	19	25	39	57
FEPB (6 thru 2), RHW and RHH (without outer covering)	0		1	1	2	3	5	8	12	16	21	33	49
	00		1	1	1	3	5	7	10	14	18	29	41
	000		1	1	1	2	4	6	9	12	15	24	35
	0000			1	1	1	3	5	7	10	13	20	29
	MCM												
	250			1	1	1	2	4	6	8	10	16	23
	300			1	1	1	2	3	5	7	9	14	20
	350				1	1	1	3	4	6	8	12	18
	400				1	1	1	2	4	5	7	11	16
	500				1	1	1	1	3	4	6	9	14
	600					1	1	1	3	4	5	7	11
	700					1	1	1	2	3	4	7	10
	750					1	1	1	2	3	4	6	9

5. Motor Protection

a. *Conductor size*

(1) Single motor (NEC® 430-22a). " ... Branch-circuit conductors supplying a single motor shall have an ampacity not less than 125 percent of the motor full-load current rating ..."

(2) Several Motors (NEC® 430-24). "Conductors supplying two or more motors shall have an ampacity equal to the sum of the full-load current rating of all the motors plus 25 percent of the highest rated motor in the group. ..."

b. *Short Circuit Protection* (NEC® 430-52). "The motor branch-circuit, short-circuit and ground-fault protective device shall be capable of carrying the starting current of the motor. A protective device having a rating or setting not exceeding the value calculated according to the values given in Table 430-152 shall be permitted."

"Exception No. 1. Where the values for branch-circuit short-circuit and ground-fault protective devices determined by Table 430-152 do not correspond to the standard sizes or ratings of fuses, non-adjustable circuit breakers, for thermal protective devices, or possible settings of adjustable circuit breakers adequate to carry the load, the next higher size, rating, or setting shall be permitted."

"An instantaneous trip circuit breaker shall be used only if adjustable, and if part of combination controller having motor overload and also short-circuit and ground-fault protection in each conductor. A motor short-circuit protector shall be permitted in lieu of devices listed in Table 430-152 if the motor short-circuit protector is part of a combination controller having both motor overload protection and short-circuit and ground-fault protection in each conductor and if it will operate at not more than 1300 percent of full-load motor current. An instantaneous trip circuit breaker or motor short-circuit protector shall be used only as part of a combination motor controller which provides coordinated motor branch-circuit overload and short-circuit and ground-fault protection."

c. *Overload Protection* (NEC® 430-32(a)(1)). "Each continuous-duty motor rated more than one horsepower shall be protected against overload. (1) A separate overload device that is responsive to motor current. This device shall be selected to trip or rated no more than the following percent of the motor nameplate full-load current rating.

Motors with a marked service factor not less than 1.15	125%
Motors with a marked temperature rise not over 40°C	125%
All other motors	115%

Table 430-7(b) Locked-rotor Indicating Code Letters		
Code Letter		**Kilovolt-Amperes per Horsepower with Locked Rotor**
A	. . . .	0 - 3.14
B	. . . .	3.15 - 3.54
C	. . . .	3.55 - 3.99
D	. . . .	4.0 - 4.49
E	. . . .	4.5 - 4.99
F	. . . .	5.0 - 5.59
G	. . . .	5.6 - 6.29
H	. . . .	6.3 - 7.09
J	. . . .	7.1 - 7.99
K	. . . .	8.0 - 8.99
L	. . . .	9.0 - 9.99
M	. . . .	10.0 - 11.19
N	. . . .	11.2 - 12.49
P	. . . .	12.5 - 13.99
R	. . . .	14.0 - 15.99
S	. . . .	16.0 - 17.99
T	. . . .	18.0 - 19.99
U	. . . .	20.0 - 22.39
V	. . . .	22.4 - and up

Table 430-148 Full-load Current§ in Amperes for Single-phase Alternating-current Motors			
hp	115V	200V‡	230V
1/6	4.4	2.5	2.2
1/4	5.8	3.3	2.9
1/3	7.2	4.1	3.6
1/2	9.8	5.6	4.9
3/4	13.8	7.9	6.9
1	16	9.2	8
1.5	20	11.5	10
2	24	13.8	12
3	34	19.6	17
5	56	32.2	28
7.5	80	46	40
10	100	57.5	50

§These values of full-load current are for motors running at usual speeds and motors with normal torque characteristics. Motors built for especially low speeds or high torques may have higher full-load currents, and multispeed motors will have full-load current varying with speed, in which case the nameplate current rating shall be used.

The voltages listed are rated motor voltages. The currents listed shall be permitted for system voltage ranges of 110 to 120, 191 to 218, and 220 to 240 V.

‡The 200 V column has been added by the author using the instruction: To obtain full-load currents of 208 V and 200 V motors, increase corresponding 230 V motor full-load currents by 10 and 15 percent, respectively.

Table 430-150

Full-load Current§ for Three-phase Alternating-current Motors

hp	Induction Type Squirrel-cage and Wound-rotor Amperes						Synchronous Type Unity Power Factor† Amperes				
	115V	200V‡	230V	460V	575V	2300V	200V‡	230V	460V	575V	2300V
1/2	4	2.3	2	1	.8						
3/4	5.6	3.22	2.8	1.4	1.1						
1	7.2	4.14	3.6	1.8	1.4						
1.5	10.4	5.98	5.2	2.6	2.1						
2	13.6	7.82	6.8	3.4	2.7						
3		11	9.6	4.8	3.9						
5		17.5	15.2	7.6	6.1						
7.5		25.3	22	11	9						
10		32.2	28	14	11						
15		48.3	42	21	17						
20		62.1	54	27	22						
25		78.2	68	34	27		61	53	26	21	
30		92	80	40	32		72	63	32	26	
40		120	104	52	41		95	83	41	33	
50		150	130	65	52		120	104	52	42	
60		177	154	77	62	16	141	123	61	49	12
75		221	192	96	77	20	178	155	78	62	15
100		285	248	124	99	26	232	202	101	81	20
125		359	312	156	125	31	291	253	126	101	25
150		414	360	180	144	37	347	302	151	121	30
200			480	240	192	49	400	460	201	161	40

‡The 200 V columns have been added by the author using the instruction: To obtain full-load currents of 208 V and 200 V motors, increase corresponding 230 V motor full-load currents by 10 and 15 percent, respectively.

§These values of full-load current are for motors running at speeds usual for belted motors and motors with normal torque characteristics. Motors built for especially low speeds or high torques may require more running current, and multispeed motors will have full-load current varying with speed, in which case the nameplate current rating shall be used.

†For 90 and 80 percent power factors, the above figures shall be multiplied by 1.1 and 1.25, respectively.

The voltages listed are rated motor voltages. The currents listed shall be permitted for system voltage ranges of 110 to 120, 191 to 218, 220 to 240, 440 to 480, 550 to 600 V.

Table 430-152

Maximum Rating or Setting of Motor Branch-circuit
Short-circuit and Ground-fault Protective Devices

Type of Motor	Percent of Full-load Current			
	Nontime Delay Fuse	Dual Element (Time-delay) Fuse	Instantaneous Trip Breaker	§Inverse Time Breaker
Single-phase, all types				
No code letter	300	175	700	250
All ac single-phase and polyphase squirrel-cage and synchronous motor† with full-voltage, resistor or reactor starting:				
No code letter	300	175	700	250
Code letter F to V.	300	175	700	250
Code letter B to E.	250	175	700	200
Code letter A	150	150	700	150
All ac squirrel-cage and synchronous motors† with autotransformer starting:				
Not more than 30 amps				
No code letter	250	175	700	200
More than 30 amps				
No code letter	200	175	700	200
Code letter F to V	250	175	700	200
Code letter B to E	200	175	700	200
Code letter A.	150	150	700	150
High-reactance squirrel-cage				
Not more than 30 amps				
No code letter	250	175	700	250
More than 30 amps				
No code letter	200	175	700	200
Wound-rotor				
No code letter	150	150	700	150
Direct-current (constant voltage)				
No more than 50 hp				
No code letter	150	150	250	150
More than 50 hp				
No code letter	150	150	175	150

For explanation of Code Letter Marking, see Table 430-7(b).

For certain exceptions to the values specified, see Sections 430-52 through 430-54.

§The values given in the last column also cover the ratings of nonadjustable inverse time types of circuit breakers that may be modified as in Section 430-52.

†Synchronous motors of the low-torque, low-speed type (usually 450 rpm or lower), such as are used to drive reciprocating compressors, pumps, etc. that start unloaded, do not require a fuse rating or circuit-breaker setting in excess of 200 percent of full-load current.

6. Transformer Protection (600 V or less) (NEC® 450-3)

Overcurrent protection. "Overcurrent Protection. ... as used in this section, the word *transformer* shall mean a transformer or polyphase bank of two or three single-phase transformers operating as a unit..."

(1) Primary. Each transformer 600 V or less shall be protected by an individual overcurrent device on the primary side, rated or set at not more than 125 percent of the rated primary current of the transformer. (See NEC® for exceptions.)

(2) Primary and Secondary. A transformer 600 V or less having an overcurrent device on the secondary side rated or set at not more than 125 percent of the rated secondary current of the transformer shall not be required to have an individual overcurrent device on the primary side if the primary feeder overcurrent device is rated or set at a circuit value not more than 250 percent of the rated primary current of the transformers.

A transformer 600 V or less, equipped with coordinated thermal overload protection by the manufacturer and arranged to interrupt the primary current, shall not be required to have an individual overcurrent device on the primary side if the primary feeder overcurrent device is rated or set at a current value not more than 6 times the rated current of the transformer for transformers having not more than 6 percent impedance, and not more than 4 times the rated current of the transformer for transformers having more than 6 but not more than 10 percent impedance. (See NEC® for exceptions.)"

APPENDIX C

CONVERSION CONSTANTS

To Convert from	to	Multiply by
angular velocity		
rev/min	rad/s	$2\pi/60 = 1/9.5493$
hp	kW	0.746
length 1 m	ft	3.281 (exact)
1 m	1 in	39.37 (exact)
mass		
1 kg	slug	0.0685
1 kg	lb mass	2.205
Force		
Newton	lb	0.225
Newton	poundals	7.23
Energy		
joule (watt-second)	lb-ft	0.738
Newton-meter		
Magnetic flux (Wb)	Maxwells (line)	10^8
Magnetic flux density		
teslas	gauss	10,000
teslas	kilolines/in^2	64.5
Magnetizing		
1 At/m	At/in	0.0254
1 At/m	oersted	0.0126
Torque		
N·m	lb·ft	0.738
lb·ft	N·m	1.355 818 0

APPENDIX D

CONSTANTS

permeability of free space $\mu_o = 4\pi\times10^{-7}$ weber/ampere-turn meter

permittivity (capacitivity) of free space

$\epsilon_o = 8.854\ 185\times10^{-12}$ coulomb^2/Newton-meter^2

also, $\epsilon_o = \dfrac{1}{\mu_o C^2}$ where $C = 2.997\ 925\times10^8$ is the velocity of
light from NASA-7012

acceleration of gravity $g = 9.807$ m/sec^2

REFERENCES
AND READINGS

A. General Machinery Texts

(1)   B. Adkins, *The General Theory of Electrical Machines*. New York: John Wiley & Sons, 1957.

(2)   L. R. Anderson, *Electric Machines and Transformers*. Reston, Va.: Reston Publishing Co. Inc., 1981.

(3)   J. C. Andreas, *Energy-Efficient Electric Motors*. New York: Marcel Dekker, 1982.

(4)   D. Brown and E. P. Hamilton, III, *Electromechanical Energy Conversion*. New York: Macmillan Publishing Co., 1984.

(5)   C. C. Carr, *Electric Machinery*. New York: John Wiley & Sons, 1958

(6)   C. R. Chapman, *Electromechanical Energy Conversion*. New York: Blaisdell Publishing Co., 1965.

(7)   S. J. Chapman, *Electric Machinery Fundamentals*. New York: McGraw-Hill Book Co., 1985.

(8)   C. D. Crosno, *Fundamentals of Electromechanical Conversion*. New York: Harcourt, Brace & World, Inc., 1968.

(9)   V. Del Toro, *Electromechanical Devices for Energy Conversion and Control Systems*. Englewood Cliffs, N. J.: Prentice-Hall, Inc., 1968.

(10) J. D. Edwards, *Electrical Machines*. Aylesburg, Buckshire, England: International Textbook Co. Ltd., 1973.

(11) O. I. Elgerd, *Electric Energy Systems Theory*, 2nd ed. New York: McGraw-Hill Book Co., 1982.

(12) A. J. Ellison, *Electromechanical Energy Conversion*. New York: Reinhold Publishing Co., 1965.

(13) R. H. Engelmann, *Static and Rotating Electromagnetic Devices*. New York: Marcel Dekker, Inc., 1982.

(14)   A. E. Fitzgerald, C. Kingsley, Jr., and S. D. Umans, *Electric Machinery*, 4th ed. New York: McGraw-Hill Book Co., 1983; (A) 3rd ed., 1971. [An abridgement of the 2nd ed.]; (B) 2nd ed. 1961. [A classic]; (C) 1st ed., 1952.

(15)   H. W. Gingrich, *Electrical Machinery Transformers and Control*. Englewood Cliffs, N. J.: Prentice-Hall, Inc., 1979.

(16)   V. Gourishankar and D. H. Kelly, *Electromechanical Energy Conversion*, 2nd ed. New York: Intext Educational Publishing, 1973.

(17)   J. Hindmarsh, *Electrical Machines and Their Applications*, 2nd ed. Oxford, England: Pergamon Press, 1970.

(18)   H. E. Koenig and W. A. Blackwell, *Electromechanical System Theory*. New York: McGraw-Hill Book Co., 1961.

(19)   Y. H. Ku, *Electric Energy Conversion*. New York: The Ronald Press Co., 1959.

(20)   V. Jones, *The Unified Theory of Electrical Machines*. New York: Plenum Press, 1967.

(21)   E. Levi and M. Panzer, *Electromechanical Power Conversion*. New York: McGraw-Hill Book Co., 1966.

(22)   M. Liwschitz-Garik and C. C. Whipple, *Electric Machinery*. Vol I, DC Machines, New York: D. Van Nostrand Co, 1946. Vol. II AC Machines, D. Van Nostrand Co., 1946.

(23)   M. Liwschitz-Garik and R. T. Weil, *DC and AC Machines*. New York: D. Van Nostrand Co., 1952.

(24)   T. C. Lloyd, *Electric Motors and Their Applications*. New York: Wiley Interscience, 1969.

(25)   W. V. Lyon, *Transient Analysis of Alternating-Current Machinery*. New York: John Wiley & Sons, 1954.

(26)   V. E. Mablekos, *Electric Machinery Theory for Power Engineers*. New York: Harper & Row, Publishers, 1980.

(27)   H. Majmudar, *Electromechanical Energy Converters*. Boston: Allyn & Bacon, Inc., 1965.

(28)   L. W. Matsch, *Electromagnetic and Electromechanical Machines*, 2nd ed. New York: IEP, 1977.

(29)   G. McPherson, *An Introduction to Electrical Machines and Transformers*. New York: John Wiley & Sons, 1981.

(30)   J. Meisel, *Principles of Electromechanical Energy Conversion*. New York: McGraw-Hill Book Co., 1966.

(31)   S. A. Nasar, *Electromechanical Energy Conversion Devices and Systems*. Englewood Cliffs, N. J.: Prentice-Hall, Inc., 1970.

(32)   S. A. Nasar, *Electric Machines and Transformers*. New York: Macmillan Publishing Co., 1984.

(33)   S. A. Nasar and L. E. Unneweher, *Electromechanics and Electric Machines*, 2nd ed. New York: John Wiley & Sons, 1983.

(34)   D. V. Richardson, *Rotating Electrical Machinery and Transformer Technology*, 2nd ed. Reston, Va.: Reston Publishing Co. Inc., 1982.

(35)   M. S. Sarma, *Electric Machines*, Dubuque, Iowa: Wm C. Brown Publishers, 1985.

(36)   D. R. Schoultz, *Electric Motors in Industry*. New York: John Wiley & Sons, 1942.

(37) S. Seeley, *Electromechanical Energy Conversion*. New York: McGraw-Hill Book Co., 1962.

(38) H. J. Skilling, *Electromechanics*. New York: John Wiley & Sons, 1962.

(39) G. R. Slemon and A. Straughen, *Electric Machines*. Reading, Mass: Addison-Wesley Publishing Co., 1980.

(40) R. Stein and W. T. Hunt, Jr., *Electrical Power System Components: Transformers and Rotating Machines*. New York: Van Nostrand Reinhold Co., 1979.

(41) G. J. Thaler and M. L. Wilcox, *Electric Machines: Dynamic and Steady State*. New York: John Wiley & Sons, 1966.

(42) E. H. Wernick, ed., *Electric Motor Handbook*. London: McGraw-Hill Book (UK) Co., 1978.

(43) D. C. White and H. H. Woodson, *Electromechanical Energy Conversion*. New York: John Wiley & Sons, 1959.

(44) H. H. Woodson and J. R. Melcher, *Electromechanical Dynamics. Part I: Discrete Systems; Part II: Fields, Forces, and Motion*. New York: John Wiley & Sons, 1968.

B. dc MACHINERY TEXTS

(1) G. G. Blalock, *Direct-Current Machinery*. New York: McGraw-Hill Book Co., 1947.

(2) Electro-Craft Corp., *DC Motors, Speed Controls, Servo Systems*, 3rd ed. Hopkins, Minn.: Electro-Craft Corp., 1975.

(3) R. G. Kloeffler, R. M. Kerchner, and J. L. Brenneman, *Direct-Current Machinery*. New York: Macmillan Publishing Co., 1949.

(4) A. S. Langsdorf, *Principles of Direct-Current Machines*, 5th ed. New York: McGraw-Hill Book Co., 1940. (This book includes a considerable amount of magnetic theory.); 6th ed. 1959. (The magnetic theory was left out of the 6th ed.)

(5) A. F. Puchstein, *The Design of Small Direct-Current Motors*. New York: John Wiley & Sons, 1961.

(6) C. S. Siskind, *Direct-Current Machinery*. New York: McGraw-Hill Book Co., 1952.

C. ac MACHINERY TEXTS

(1) B. F. Bailey and J. S. Gault, *Alternating Current Machinery*. New York: McGraw-Hill Book Co., 1951.

(2) L. V. Bewley, *Alternating Current Machinery*. New York: Macmillan Publishing Co., 1949.

(3) J. M. Bryant and E. W. Johnson, *Alternating Current Machinery*. New York: McGraw-Hill Book Co., 1935.

(4) A. E. Langsdorf, *Theory of Alternating Current Machinery*, 2nd ed. New York: McGraw-Hill Book Co., 1955.

(5) R. R. Lawrence, *Principles of Alternating Current Machinery*, 3rd ed. New York: McGraw-Hill Book Co., 1940.

(6) W. A. Lewis, "The Principles of Synchronous Machinery." Chicago: Illinois Institute of Technology, 1959. (Unpublished Manuscript)

(7)   M. Liwschitz-Garik and C. C. Whipple, *Electric Machinery.* Vol. II, AC Machines. New York: D. Van Nostrand Co., 1946.

(8)   T. C. McFarland, *Alternating Current Machines.* New York: D. Van Nostrand Co., 1948.

(9)   G. V. Mueller, *Alternating Current Machines.* New York: McGraw-Hill Book Co., 1952.

(10)   A. F. Puchstein, T. C. Lloyd, and A. G. Conrad, *Alternating Current Machines,* 4th ed. New York: John Wiley & Sons, 1954.

(11)   J. Rosenblatt and M. H. Friedman, *Alternating Current Machinery.* New York: McGraw-Hill Book Co., 1963. (A technology text)

(12)   J. G. Tarboux, *Alternating Current Machinery.* Scranton, Pa.: International Textbook Co., 1947.

D. SYNCHRONOUS MACHINES

(1)   C. Concordia, *Synchronous Machines, Theory and Performance.* New York: John Wiley & Sons, 1957.

(2)   M. S. Sarma, *Synchronous Machines.* New York: Gordon and Breach, 1979.

E. INDUCTION MACHINES

(1)   P. L. Algers, *The Nature of Induction Machines.* New York: Gordon and Breach, 1965.

F. FRACTIONAL HORSEPOWER MOTORS

(1)   Bodine Fractional-horsepower Motor Handbook, Bodine Electric Co., 2500 West Braley Place, Chicago, Ill., 1959.

(2)   C. G. Veinott, *Fractional Horsepower Electric Motors,* 2nd ed. New York: McGraw-Hill Book Co., 1948.

(3)   C. G. Veinott, *Theory and Design of Small Induction Motors.* New York: McGraw-Hill Book Co., 1959.

G. FEEDBACK AND CONTROL DEVICES

(1)   W. R. Ahrendt, *Ahrendt and Savant, Servomechanism Practice,* 2nd ed. New York: McGraw-Hill Book Co., Inc., 1960.

(2)   Blackburn, *Components Handbook, MIT Radiation Laboratory Series,* Vol. 17. New York: McGraw-Hill Book Co., 1949.

(3)   J. E. Gibson and F. B. Tuteur, *Control System Components.* New York: McGraw-Hill Book Co., 1958.

(4)   H. M. James, N. B. Nichols, and R. S. Phillips, *Theory of Servomechanics.* New York: McGraw-Hill Book Co., 1947.

(5)   J. M. Pestarini, *Metadyne Statics.* New York: John Wiley & Sons, 1952.

(6)   A. Tustin, *Direct Current Machines for Control Systems.* New York: Macmillan Publishing Co., 1952.

(7)   *United States Navy Synchros, Description and Operation.* Supt. of Documents, U. S. Government Printing Office, Washington, D. C., OP 1303, 15 Apr 1958.

(8)   Van Valkenburgh, Noogen and Neville, Inc., *Basic Synchros and Servomechanisms*. New York: John F. Rider Publisher, Inc., 1965.

H. MOTOR CONTROLS

(1)   R. L. McIntire, *A-C Motor Control Fundamental*. New York: McGraw-Hill Book Co., 1960.

I. DIAKOPTICS

(1)   L. V. Bewley, *Tensor Analysis of Circuits and Machines*. New York: The Ronald Press Co., 1961.

(2)   H. H. Happ, *Diakoptics and Networks*. New York: Academic Press, 1971.

(3)   G. Kron, *Equivalent Circuits of Electric Machinery*. New York: John Wiley & Sons, 1951.

(4)   G. Kron, *Tensor Analysis of Networks*. New York: John Wiley & Sons, 1939. [This is the unabridged version.]

(5)   G. Kron, *Tensor for Circuits*, 2nd ed. New York: Dover Publications, Inc., 1961.

J. TRANSFORMERS AND MAGNETIC CIRCUITS

(1)   R. L. Bean, et al., *Transformers for the Electric Power Industry*. New York: McGraw-Hill Book Co., 1959.

(2)   L. F. Blume, et al., *Transformer Engineering*, 2nd ed. New York: John Wiley & Sons, 1951. GE engineers. A reprint of chapters 2, 3, and 4 are available as: L. F. Blume and A. Boyajean, Transformer Connections, General Electric Co., Schenectady, N.Y., GET-2D, Jan. 1947.

(3)   W. R. Brown, Jr., *Magnetostatic Principles in Ferromagnetism*. New York: Interscience Publishing, Inc., 1962.

(4)   General Electric Co., *Permanent Magnet Manual*. PM200, Edmore, Mich. 48829, 1963.

(5)   J. B. Gibbs, *Transformer Principles and Practice*, 2nd ed. New York: McGraw-Hill Book Co., 1950.

(6)   W. T. Hunt, Jr., and R. Stein, *Static Electromagnetic Devices*. Boston: Allyn & Bacon 1963.

(7)   P. Mathews, *Protective Current Transformers and Circuits*. New York: Macmillan Publishing Co., 1955.

(8)   L. W. McKeehan, *Magnets*. New Jersey: Van Nostrand, 1967.

(9)   R. Lee, *Electronic Transformers and Circuits*. New York: John Wiley & Sons, 194 (Written by a Westinghouse consultant. This book included design details for rectifie filament, plate, pulse, and control transformers.)

(10)   L. W. Matsch, *Capacitors, Magnetic Circuits, and Transformers*. Englewood Clif N.J.: Prentice-Hall, Inc., 1964.

(11)   A. Shure, ed., *Transformers*. New York: John F. Riley Publisher, Inc., 1963. (practi presentation, 79 pages)

(12)   W. C. Sealey, *Transformers, Theory and Construction*. New York: International Text-book Co., 1946. (Description of construction and use by an Allis-Chalmers engineer.)

(13)   MIT Staff, *Magnetic Circuits and Transformers*. New York: John Wiley & Sons, 1943.

K. BATTERIES

(1)   G. W. A. Drummer, *Modern Electronic Components*. New York: Philosophical Library, 1959. (Chapter 16, Batteries and Accumulators. A good 8-page summary of batteries by a British author.)

(2)   G. W. Vinal, *Primary Batteries*. New York: John Wiley & Sons, 1950.

(3)   G. W. Vinal, *Storage Batteries*, 3rd ed. New York: John Wiley & Sons, 1940.

L. RECTIFIERS

(1)   J. M. Shaeffer, *Rectifier Circuits*. New York: John Wiley and Sons, 1965.

M. SERVICE COURSE TEXTS

(1)   V. Del Toro, *Electrical Engineering Fundamentals*. Englewood Cliffs, N. J.: Prentice-Hall, Inc., 1972.

(2)   A. E. Fitzgerald, D. E. Higginbotham, and A. Grabel, *Basic Electrical Engineering*, 5th ed. New York: McGraw-Hill Book Co., 1982.

N. HISTORY

(1)   E. T. Canby, *A History of Electricity*. New York: Hawthorne Books, 1963.

(2)   P. Dunsheath, *A History of Electrical Engineering*. London: Faber and Faber, 1962.

(3)   J. K. Finch, *The Story of Engineering*. Garden City, N.Y.: Anchor Books, Doubleday & Co., Inc., 1960.

(4)   J. A. Fleming, *Fifty Years of Electricity, The Memories of an Electrical Engineer*. London & New York: The Wireless Press, Ltd., 1921. [History of telegraph, telephone, electrical generation, heating, measurements, etc. - by an Englishman.]

(5)   J. W. Hammond, *Men and Volts, The Story of the General Electric Co*. New York: J. B. Lippincott Co., 1941.

(6)   P. W. Kingsford, *Electrical Engineering, A History of the Men and the Ideas*. New York: St. Martin's Press, 1970.

(7)   F. A. Lewis, *The Incandescent Light,* revised ed. New York: Shorewood Publishing, Inc., 1961. The story of Thomas Edison and the first incandescent lamps.

(8)   H. W. Meyer, *A History of Electricity and Magnetism*. Norwalk, Conn.: Bundy Library, 1972.

(9)   A. P Morgan, *The Pageant of Electricity*. Appleton-Century Co. (History of equipment development.)

(10)   Brother Potamina (O'Reilly) and James J. Walsh, *Makers of Electricity*. New York: Fordham University Press, 1909. (Biographies of Gilbert, Galvani, Volta, Ampere, etc.)

(11)   S. Rapport and H. Wright, *Engineering*. New York: University Press, 1963. (A collection of historic sketches which include stories about Leonardo da Vinci, James Watt, Wire rope, Brooklyn Bridge, Tesla, etc.)

(12)   A. Still, *Soul of Amber*. New York: Murray Hill Books, Inc., 1944.

(13)   H. B. Walters, *Nikola Tesla*. New York: Thomas Y. Crowell Co., 1961. An interestingly written biography.

(14)   D. O. Woodbury, *Beloved Scientist*. New York: Whittlesey House of McGraw-Hill Book Co., 1944. (Biography of Elihu Thomson.)

O. MISCELLANEOUS

(1)   Bodine Electric Co., *Motorgram*, Bodine Electric Company, Chicago, Ill. (Published bimonthly).

(2)   General Electric, "Relaying for Industrial Electrical Power Systems and Equipment," Bulletin GET-7203A.

(3)   W. H. Beyer, *Standard Mathematical Tables*, 27th ed. Boca Raton, Florida: CRC Press, Inc., 1985.

(4)   C. R. Wylie and L. C. Barrett, *Advanced Engineering Mathematics*, 5th ed. New York: McGraw-Hill Book Co., 1982.

(5)   Standards.

a. ANSI C50.10-1977. General Requirements of Synchronous Machines (See Tests, Sect. 9.2)

b. ANSI C50.13-1977. Requirements for Cylindrical Rotor Synchronous Machines (See Tests, Sect. 10)

c. ANSI C50.14-1977. Requirements for Combustion Gas Turbine Driven Cylindrical Rotor Synchronous Machines.

d. ANSI C57.12.00-1980. "General requirements for liquid-immersed distribution, power and regulating transformers."

e. ANSI C57.12.01-1979. "General requirements for dry-type distribution and power transformers."

f. ANSI C57.12.90-1980. "Test code for liquid-immersed distribution power transformers..."

g. IEEE Std 58-1978. IEEE Standard Induction Motor Letter Symbols.

h. IEEE Std 100-1977. IEEE Standard Dictionary of Electrical & Electronic Terms. For machinery topics see titles such as (1) phasor diagram (synchronous machine), Potier reactance, synchronous generator.

i. C57.12.105-1978. "Guide for application of transformer connections in three-phase distribution systems."

j. IEEE Std. 112-1978. Standard Test Procedures for Polyphase Induction Motors and Generators.

k. IEEE Std 115-1965. "Test Procedures for Synchronous Machines" (See section 4.01 for load excitation and voltage regulation.)

INDEX